COMPOSITE MECHANISMS

Yan Zhenying

Order this book online at www.trafford.com
or email orders@trafford.com

Most Trafford titles are also available at major online book retailers.

Printed in the United States of America.

ISBN: 978-1-4269-5178-7 (sc)
ISBN: 978-1-4269-5270-8 (e)

Trafford rev. 10/03/2011

 www.trafford.com

North America & international
toll-free: 1 888 232 4444 (USA & Canada)
phone: 250 383 6864 ♦ fax: 812 355 4082

AFFECTIONATELY INSCRIBED

TO

THE BANKER MR. AND MRS. QU,JIGANG

THE DOCTOR MR. AND MRS.YAN,QINGKANG

Preface

Pioneers have devoted themselves to creating new machines for hundreds of years. How to improve and develop the functions and performance of simple mechanisms is the fascinating ideal among scholars and engineers in the mechanical industry. Vast amount of investigations, reports, papers, and real machines have been shown vividly. *Composite Mechanisms* is one of the very category of these kinds. They are from basic mechanisms, such as linkages, gears, cams, and so on. Once these simplest mechanisms are composed, they will have better qualities and performance than they had before. This is the charm of *Composition*, the soul of it. This book deals with an introduction, geared linkages, cam-link mechanisms, cam-gear mechanisms and the others, general review of composite mechanisms, and appendixes.

The introduction provides the subjects studied, the purpose of studying, the means to be used, and the historical background concerned.

Geared linkages, especially the geared five-bar linkages, are better than conventional four-bar linkages at the function, path, and motion generations. However, the marvelous shapes of the path curve, the velocities, and accelerations created by the moving particle (i.e. any point on the member of the mechanism) are under the designer's command. Geared linkages can create more precise positions. The geared four-bar linkages is still a focus of geared linkages for its broad applications, such as the cycloidal crank mechanism, etc. Besides, the six-bar and three-bar linkages are available also.

Cam-link mechanisms might be the finest generators for the function , path and motion if wear and balancing problems were not to be considered. The cam and linkage connected either in series or in parallel will be designed with approaches, geometrically and algebraically. The fundamental principles of robot (i.e. the two-link manipulator) will be discussed fully.

Cam-gear mechanisms, Geneva wheels, and chains can be made into different kinds of intermittent motions and various kinematic characteristics. The composition of these mechanisms is another field of creation.

The general review deals with the force analysis on, the advantages and disadvantages of composite mechanisms. A model example for output stroke extension is discussed minutely in this part.

Appendixes cover many subjects, such as the Sylvester's dialytic elimination, dead center positions, the specific overall acceleration equation for mechanisms connected in series, and the pressure angle…

All the description will let you know what it is. Through examples, single or complex, you may find the ways to follow and the new machines to create.

Knowledge of fundamental algebra, geometry, differential and integral calculus, matrix, kinematics, and mechanisms is needed for studying. However, the bumper harvest will be obtained after hard and elaborate working. Everyone can be a good planter and can secure many brilliant fruits if they are interested in it and do their best. Many blanks remain to be filled and conquered. Success always belongs to those who work diligently.

I am greatly indebted to many friends and colleagues, especially the former Presidents Jiang Kwhua,Chen Liguo, Huang Wenzhi, Mectthanical Engineering Department Chairman Deng Zuan. Professor Yuan Jiayi of Wuhan Institute of Technology; the Dean Xia Zhongxian of Jilin Polytechnic University for their helps and supports. Also I would like to acknowledge my wife Qu Yanru and daughter-in-law Zhang Juanman for their encouragement and helpfulness.

Finally this book would not have been underetaken without the brilliant works and excellent results of the most outstanding scholars who are listed (and those not listed) in reference at the end of the book. I am just standing on the shoulders of Titans'. I cordially salute them forever.

Yan Zhenying

List of Symbols

A _____ coefficient , area

A_0 _____ specific acceleration of two component mechanisms connected in series

A_I _____ specific acceleration of the first component mechanism

A_{II} _____ specific acceleration of the second component mechanism

a _____ coefficient , length, linear acceleration, distance of axis to axis
B _____ coefficient
b _____ coefficient, length
C _____ coefficient, center, distance of axis to axis
c _____ coefficient, center, distance of axis to axis, length
D _____ coefficient, diameter
d _____ diameter, arm of couple, distance of axis to axis, length
E _____ coefficient, error
e _____ Napierian logarithmic base, length
F _____ coefficient, force, number of degrees of freedom
f _____ length

i _____ $\sqrt{-1}$

j _____ ordinal
K _____ coefficient, ratio, unknown
k _____ ratio, ordinal
L _____ ratio, lead, length
l _____ coefficient, unknown, length
M _____ ratio, moment
MA _____ mechanical advantage
m _____ ratio, coefficient, unknown, module, radius of translation
N _____ ratio, normal force, numbers of teeth, grooves, and pitches
n _____ transmission ratio, number, unknown

n_0 _____ total transmission ratio

n_I _____ transmission ratio of the first component mechanism

n_{II} _____ transmission ratio of the second component mechanism

P _____ force, instantaneous center, pitch
p _____ coefficient, number
Q _____ instantaneous center
q _____ distance between instantaneous center of two side links to related rotating center on a four-bar linkage
R _____ radius, ratio, coefficient, length, scale factor
r _____ radius, length
S _____ length of stroke, circle
s _____ length of stroke, linear displacement
T _____ period

t _____ function, time
V _____ velocity
v _____ velocity
W _____ width , work

α _____ angular magnitude, position* , and displacement

α_0 _____ initial angular position

α_1 _____ first or initial angular position

α_j _____ jth angular position

α_{0j} _____ angular displacement from initial to jth position

α_{1j} _____ angular displacement from first to jth position

α_e _____ angular displacement with arm relative to frame

α_r _____ angular displacement relative to arm

α_a _____ total angular displacement relative to frame

$\beta, \gamma, \delta, \theta, \mu, \phi, \psi$ are all the same as α

ε _____ angular acceleration
λ _____ angular magnitude, unknown, ratio

ρ _____ radius of curvature, variable radius

$[\mu_{min}]$ _____ allowable minimum transmission angle

ϖ _____ angular velocity

---

* All the angular positions are measured counterclockwise from the X-axis datum in the Cartesian coordinate

Introduction

All machines are composed of simple mechanisms____gears, cams, linkages, belts, chains, Geneva wheels, and so on. For instance, an automobile is the assemblies of a power plant, power transmission, steering gear, running gear, braking system, differential, and the rest. Nearly all the assemblies are composed of the simple mechanisms listed above. Not only automobiles, but also machine tools, sewing machines, type writers, automatic assembly lines, and those assembled together for special purposes. They can be either a novel invention or a harmonic engineering system. However, these machines are not the subjects we are concerned with.

The subjects which we would study are called Composite Mechanisms___ geared linkages, cam-link mechanisms, cam-gear mechanisms, chained linkages, and their combinations.

Not all the machines come from composite mechanisms.

The purpose of studying composite mechanisms is to improve the characteristics and functions of common mechanisms. Since composite mechanisms originate from simple mechanisms; they are easy to be manufactured and assembled, and cheap in cost. Along with the reasonable design, they may be smooth, harmonic, noiseless, shockless and can be efficiently operated.

Certainly, the research of composite mechanisms should not be restricted in planar mechanisms; further more, it should not be only one degree of freedom. However, planar composite mechanisms with freedom of one degree have been studied most, consequently the achievement is great. In this book, planar composite mechanisms with one degree of freedom are to be studied mainly.

Studying mechanisms is for designing mechanisms. Design in mechanism field is called synthesis. Since analysis is the premise of synthesis; the more is analyzed, the easier is synthesized. We would do more work in analysis than that in synthesis. And the synthesis is just the dimensional synthesis*.

The synthesis of planar composite mechanism has two methods___the graphical and the analytical.

The former is based on kinematic geometry dealing with finitely separate positions and infinitesimally close positions of planar configurations or kinematic planes. It needs the knowledge of algebraic curves, differential geometry, projective geometry and so on. Therefore, it seems as a difficult job to be handled, especially for the engineers who are not familiar with theoretical mechanisms.

The latter needs iterative method with computer programming to solve the problems. It may be tedious though.   The fundamental knowledge of vector, complex numbers, matrix , differential equations, and high order algebraic equations is required only.   It is welcome for its simplicity and popularity.   Besides, being an auxiliary means of analytical method, primary graphical approaches which do not belong to the category of such puzzled classical geometry is adopted. It just involves the basic geometrical axioms and theorems. Owing to its simplicity and clarity, it can guide the way for problem solving and check the analytical computing results. The analytical method with the graphical means as its cooperator will play an important role in the study of composite mechanisms in this text.

Finally, We should know that the composition of simple mechanisms was a reality as ancient as, before Christ in China, the south-pointing chariot (composition of different gear trains) shown in Fig.0.1; and in eighteenth century in England, the Watt's steam engine (composition of epicyclic gear train and linkages) shown in Fig.0.2.

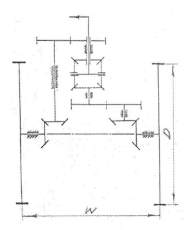

Figure 0.1 South-pointing chariot configuration, before Christ in China, gear ratios all in 1:1. *D*—wheel diameter ,*W*—track width.

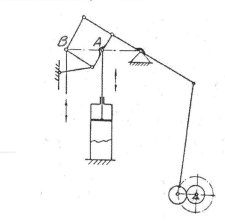

Figure 0.2   Watt's steam engine, 1784 in England. The coupler point *A* of Watt's figure 8 shaped mechanism partly describes an approximate straight line. And this vertical segment was used for piston rod guidance. By means of a pantograph point *B* was used for condensation pump in the same way.

*There are three phases of synthesis: (1) type synthesis ; (2) number synthesis ;(3)dimensional synthesis.

CONTENTS

Part IV
General Review of Composite Mechanisms

Part I   Geared Linkages

1   Geared Five-Bar Linkages

1.1   Two-Gear Five-Bar Linkage

1.1.1   How the Two-Gear Five-Bar Linkage Comes From and Its Coupler Curves

While the planar four-bar linkage (i.e. commonly called four-bar linkage) has been studied so widely and prolonged so long; the five-bar, six-bar, and even more bars linkages are being studied by many scholars all over the world. Since the five-bar linkage is next to four-bar linkage, it would be the focus of these objects naturally. As the planar five-bar linkage has inherently two degrees of freedom as shown in Fig.1.1a, then a pair of meshing gears should be attached to the corresponding links. However, a two-gear five-bar linkage with one degree of freedom is produced in Fig.1.1b.

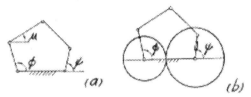

Figure 1.1 (a) Five-bar linkage, two degrees of freedom. Once ϕ and μ are set, ψ will be obtained; (b) two-gear five-bar linkage, one degree of freedom. Setting ϕ, obtaining ψ.

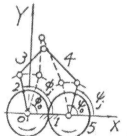

.   Figure 1.2 Two-gear five-bar linkage

A   typical two-gear five-bar linkage is shown in Fig.1.2. The lengths of the

five-bar are L_1, L_2, L_3, L_4, and L_5 respectively. The driving and driven members'

phase angles in the jth position are ϕ_j, ψ_j ; in the initial position, ϕ_0, ψ_0 respectively.

All the angles are measured from X-axis counterclockwise.

The driving and driven members' angular displacements are ϕ_{0j}, ψ_{0j} respectively.

Their relationship is as follows:

$$\psi_{0j} = n\phi_{0j} \tag{1.1}$$

where $\psi_{0j} = \psi_j - \psi_0$ $\phi_{0j} = \phi_j - \phi_0$, $n = \pm \dfrac{N_i}{N_0}$, n is called the transmission ratio of

a simple gear train, it is the reciprocal of velocity ratio. The n is positive when the input and output gears rotate in the same direction, negative if they rotate in the opposite directions. Whereas N_i and N_0 represent the numbers of teeth of input and output gears respectively. Equation (1.1) can be rewritten as

$$\psi_J - \psi_0 = n(\phi_j - \phi_0)$$

Since ϕ_0 and ψ_0 are arbitrarily chosen, and the relationship between them has relativity only, therefore setting $\phi_0 = 0$ is suitable for simplicity, we have

$$\psi_j = \psi_0 + n\phi_j \qquad\qquad (1.2)$$

Change the value of ψ_0, the magnitude and direction of n; then the configuration of the mechanism will be varied even if the dimensions of the links remain unchanged. Consequently, the path of the common joint of link 3 and 4 will be marvelous. Of course these curves are different from the curves produced by the planar four-bar linkage. They will be more astonishing if the point aside the common joint on the links is to be chosen. Now, let 30^0 be the increment of ψ_0, accompanying with the

Table 1.1 The Special Point (i.e. Common Joint) Coupler Curve Spectra of Two-Gear Five-Bar Linkages $2L_3 = 2L_4 > L_5 + L_1 + L_2$ [a] , $L_2 = r_2$ [b], $L_5 = r_5$

Type	n	L_1	L_2	L_3	L_4	L_5	d [c]	No. of Figure	Characteristics
A	$-\dfrac{1}{2}$	0.75	0.25	1.2	1.2	0.5	0	1.3	Identical curves reappear every 180^0 of output angular displacement
B	$+\dfrac{1}{2}$	1.25	0.25	1.2	1.2	0.5	0.5	1.4	Above
C	$+\dfrac{1}{2}$	0.25	0.25	1.2	1.2	0.5	0	1.5	Above. Besides, the accessible regions [d] of common joint of couplers connected together
D	-1	1.0	0.5	1.2	1.2	0.5	0	1.6	Curves are symmetrical according to $\pm\psi_0$
E	$+1$	1.25	0.5	1.2	1.2	0.5	0.25	1.7	Above. Besides, curves are symmetrical about the principle axes of their own

Notes   (a) It is set for the control of the accessible regions, categories of various curves, and avoidance of change point. All the L's and their subscripts are the lengths of links and which linkages they belong to separately.

(b) The r's are the radii of corresponding gears.

(c) The d is the diameter of idler gear. Though the configuration seems as three-gear drive, the coupler curve is similar to two-gear drive. As the idler gear merely changes the sense of rotation, so we still treat it as two-gear five-bar linkage, and is being studied here.

(d) See Appendix A.

variations of n and link lengths, a special point (i.e. common joint) coupler curve spectra will be shown in Table 1.1 and Figs 1.3 ~ 1.7

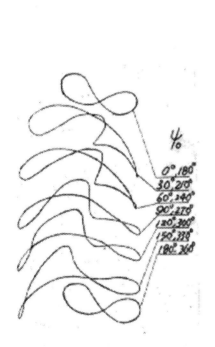

Figure 1.3 Type A

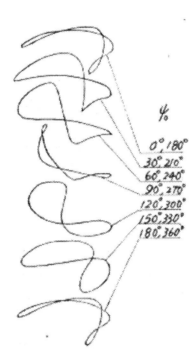

Figure 1.4 Type B

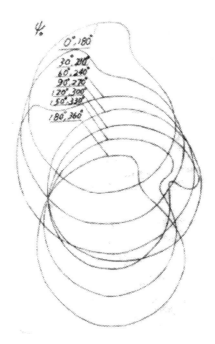

Figure 1.5   Type C

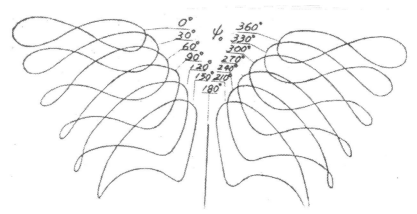

Figure 1.6   Type D

Figure 1.7 Type E

According to the above statement and the prescribed lengths of the links listed, the rules are valid consequently as follows:

(1) As for $1 \geq n \geq -1$ and $n = \pm N_i/N_0$, $1/n$ is to be the integer except zero, the curves will be the same every $2\pi n$ of output angle ; .

(2) For $n = \pm 1$, the curves will be the same every 2π of output angle. Besides, curves are symmetrical according to $\pm \psi_0$. Furthermore, for $n = +1$, they are symmetrical about the principle axes of their own. ;

(3) Let $L_2 \neq r_2$, $L_5 \neq r_5$, $L_3 \neq L_4$, and vary the numbers and the magnitudes of idler gears, then the point on links 3 and 4 aside the common joint will generate a variety of curves tremendously. ;

(4) Alike to the curves generated by planar four-bar linkages, these curves can be approximate straight lines, circular arcs. And the velocity of the moving particle may even have a nearly uniform velocity district, though it is arbitrary ordinarily. The functions of output, such as approximate uniform velocity, dwell, pilgrim motion etc., can be similar to four-bar linkages; in addition, traces may be multiply overlapped, so that the numbers and lengths of the output strokes will be various. Fig 1.8 shows a composite mechanism having four-stroke cycle with different stroke lengths.

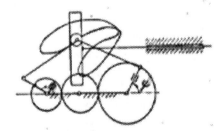

Figure 1.8   Unequal distance multi-stroke mechanism

1.1.2   Synthesized Linear Equations of Inverted Two-Gear Five-Bar Linkage

1.1.2.1   The Path and the Motion Generation (Trace of a Moving Particle and Rigid Body Guidance)

The curves studied above are marvelous though , they are generated by known mechanism configurations, so they are limited. If a definite path or motion is to be generated by the designer's own will, however, it is quite a different job. Therefore, inverting the frame of the mechanism to fulfill the designer's attempt becomes necessary.

Besides link 1 being the frame as shown above, links 2, 3, 4, or 5 each can be a frame occasionally. Since links 2 and 5 both have identical functions due to their symmetry, the same are links 3 and 4; therefore, links in each pair can be alternatively chosen as inversion.

First, taking link 5 (or 2) as the frame in Fig.1.9, a cycloidal crank mechanism appears. Point A generates a cycloid (in fact, it is a trochoid because point A is off the circumference of the rolling circle), but point B traces an arc only. The rest on links 2 and 3 may generate a lot of curves also. It is known that the cycloidal crank mechanism has been studied somewhere minutely already, and this text will introduce it appropriately later.

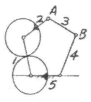

Figure 1.9   An inversion of two-gear five linkage (cycloidal crank mechanism)

Secondly, taking link 4 (or 3) as the frame, Fig.1.10, another inversion of two-gear five-bar linkage appears We call it inverted two-gear five-bar linkage just for distinction, though the cycloidal crank mechanism is also an inversion in fact. Link 2 must be relatively short, as in Fig.1.10b, otherwise this mechanism will not operate, as in Fig.1.10a.

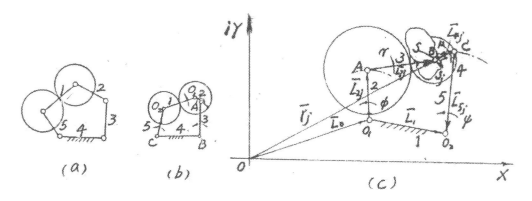

Figure 1.10 Another inversion of two-gear five-bar linkage: (a) unavailable configuration; (b) available configuration; (c) configuration in complex plane.

Path of point O_1 has to wander up and down around the circular track of point A.

It may seem that studying this mechanism is not worthwhile. However, this mechanism can be used not only for path generation but also for function generator.

So it should be investigated fully.

For discussing easier, the configuration should be placed in a complex plane, Fig.1.10c. The link numbers and revolute joint names should be recovered as before; i.e. the frame being 1, the others 2, 3, 4, and 5 arranged one by one in order, and the joints from A to C. Let points O_1 and O_2 be rotating centers of the input and output members respectively. Set point B on link 3 as the moving particle, then it will be located at S_j on path $S-S_j$ if this mechanism is in its jth position.

From origin O draw two vectors $\bar{r}_j$ and $\overline{L}_0$ to points S_j and O_1 respectively.

Let the first and second bits of $L's$ subscripts represent the ordinal number and the position of the link respectively, then two vector equations of closure can be written as follows:

$$\left.\begin{array}{l} \overline{L}_{0j} + \overline{L}_{2j} + \overline{L}_{3j} = \bar{r}_j \\ \overline{L}_{2j} + \overline{L}_{3j} + \overline{L}_{4j} + \overline{L}_{5j} = \overline{L}_{1j} \end{array}\right\}$$

$$(1.3)$$

where $\overline{L}_{0j} = \overline{L}_0, \ \overline{L}_{1j} = \overline{L}_1$

Any link vector $\overline{L}_{kj}$ (k is the ordinal number of the link) may resolve into two parts, real and imaginary, so

$$\overline{L}_{kj} = L_{kjx} + iL_{kjy}$$

in the first position

$$\overline{L}_{k1} = L_{k1x} + iL_{k1y}$$

While $\overline{L}_{k1}$ rotates an angle α_{1j} (i.e. from the first position to the jth position) counterclockwise, its resolutions L_{k1x} and L_{k1y} rotate the same amount α_{1j}, as shown in Fig.1.11. The relationship between $\overline{L}_{kj}$ and $\overline{L}_{k1}$ is as follows:

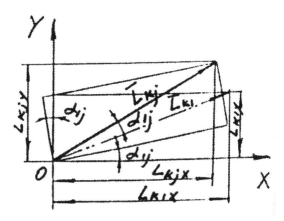

Figure 1.11 Geometrical relationship of $\overline{L}_{kj}$ and $\overline{L}_{k1}$

$$\left. \begin{array}{l} L_{kjx} = L_{k1x}\cos\alpha_{1j} - L_{k1y}\sin\alpha_{1j} \\ L_{kjy} = L_{k1x}\sin\alpha_{1j} + L_{k1y}\cos\alpha_{1j} \end{array} \right\}$$

(1.4)

besides,

$$\overline{L}_{kj} = e^{i\alpha_{1j}} \overline{L}_{k1}$$

(1.5)

where $e^{i\alpha_{1j}}$ is regarded as a rotating factor.

Now eq. (1.3) may be written as

$$\left. \begin{array}{l} \overline{L}_0 + e^{i\phi_{1j}}\overline{L}_{21} + e^{i\gamma_{1j}}\overline{L}_{31} = \overline{r}_j \\ e^{i\phi_{1j}}\overline{L}_{21} + e^{i\gamma_{1j}}\overline{L}_{31} + e^{i\mu_{1j}}\overline{L}_{41} + e^{i\psi_{1j}}\overline{L}_{51} = \overline{L}_1 \end{array} \right\}$$

(1.6)

where $\phi_{1j}, \gamma_{1j}, \mu_{1j}$ and ψ_{1j} are the angular displacements of links 2, 3, 4, and 5 from initial to jth position respectively. By eq.(1.4), it can be expressed as

$$L_{0x} + L_{21x}\cos\phi_{1j} - L_{21y}\sin\phi_{1j} + L_{31x}\cos\gamma_{1j} - L_{31y}\sin\gamma_{1j} = r_{jx}$$

(1.7-1)

$$L_{0y} + L_{21x}\sin\phi_{1j} + L_{21y}\cos\phi_{1j} + L_{31x}\sin\gamma_{1j} + L_{31y}\cos\gamma_{1j} = r_{jy}$$

(1.7-2)

$$L_{21x}\cos\phi_{1j} - L_{21y}\sin\phi_{1y} + L_{31x}\cos\gamma_{1j} - L_{31y}\sin\gamma_{1j} + L_{41X}\cos\mu_{1j} - L_{41y}\sin\mu_{1j}$$

$$+ L_{51x}\cos\psi_{1j} - L_{51y}\sin\psi_{1j} = L_{1x}$$

(1.7-3)

$$L_{21x}\sin\phi_{1j} + L_{21y}\cos\phi_{1j} + L_{31x}\sin\gamma_{1j} + L_{31y}\cos\gamma_{1j} + L_{41x}\sin\mu_{1j}$$

$$+ L_{41y}\cos\mu_{1j} + L_{51x}\sin\psi_{1j} + L_{51y}\cos\psi_{1j} = L_{1y}$$

(1.7-4)

$$\mu_{1j} = \gamma_{1j} + (\phi_{1j} - \gamma_{1j})n$$

(1.8)

where $\qquad n = \pm \dfrac{N_i}{N_0}$,

it can be proved by epicylic gear train theory.

Since ϕ_{1j}, ψ_{1j} and n being the parameters prescribed usually, and the parameter γ_{1j} arbitrarily chosen, μ_{1j} can be obtained from eq.(1.8). Then the dimensional synthesis of first order approximation of the mechanism with three finitely separated precision positions (i.e. precision points) can be presented in matrix as follows:

$$\begin{bmatrix} 0.........1.......1..........0.........0 \\ 0......e^{i\phi 12}....e^{i\gamma 12}.........0.........0 \\ 0......e^{i\phi 13}....e^{i\gamma 13}.........0.........0 \\ -1.......1.......1..........1.........1 \\ -1....e^{i\phi 12}....e^{i\gamma 12}....e^{i\mu 12}....e^{i\psi 12} \\ -1....e^{i\phi 13}....e^{i\gamma 13}....e^{i\mu 13}....e^{i\psi 13} \end{bmatrix} \begin{bmatrix} \overline{L}_1 \\ \overline{L}_{21} \\ \overline{L}_{31} \\ \overline{L}_{41} \\ \overline{L}_{51} \end{bmatrix} = \begin{bmatrix} \overline{r}_1 - \overline{L}_0 \\ \overline{r}_2 - \overline{L}_0 \\ \overline{r}_3 - \overline{L}_0 \\ 0 \\ 0 \\ 0 \end{bmatrix}$$

(1.9)

Since the angular displacement of link 3 can be arbitrarily chosen, the rigid body to be fixed on link 3 will follow the path of point B and will make rotation predetermined, so the rigid body guidance is performed. The following tables show the analyses of the path and the motion generation (i.e. trace of the moving particle and rigid body guidance) solutions, linear and non-linear, under the number of precision positions. As for the path and the motion generation of the planar four-bar linkage, see Appendix B.

Table 1.2　　Path Generation of Inverted Two-Gear Five-Bar Linkage

No. of precision positions	No. of algebraic equations	Unknowns and their numbers	Arbitrarily chosen reals and their numbers	No .of unknowns to be solved for
ϕ_{1j}, ψ_{1j}, n prescribed; linear solution				
1	4	$\overline{L}_0, \overline{L}_1, \overline{L}_{21}, \overline{L}_{31}$ $\overline{L}_{41}, \overline{L}_{51}$　　12	4 vectors of the left 8	4
2	8	Above + γ_{12}	2 vectors of the left	8

		13	$+\gamma_{12}$ 5	
3	12	Above $+\ \gamma_{13}$ 14	$\gamma_{12},\gamma_{13},$ 2	12
ϕ_{1j}, n prescribed; non-linear solution				
4	16	Above $+\gamma_{14},\psi_{12},$ ψ_{13},ψ_{14} 18	γ_{12},γ_{13} 2	16
5	20	Above $+\gamma_{15},\psi_{15}$ 20	0	20
Nothing prescribed, non-linear solution				
6	24	Above $+\gamma_{16},\psi_{16}$ $\phi_{12},\phi_{13},\phi_{14},\phi_{15}$ ϕ_{16},n 28	$n,\gamma_{12},\gamma_{13},\gamma_{14}$ 4	24
7	28	Above $+\gamma_{17},$ ψ_{17},ϕ_{17} 31	$n,\gamma_{12},\gamma_{13}$ 3	28
8	32	Above $+\gamma_{18},$ ψ_{18},ϕ_{18} 34	n,γ_{12} 2	32
9	36	Above $+\gamma_{19},$ ψ_{19},ϕ_{19} 37	n 1	36
10	40	Above $+\gamma_{1\ 10}$ $\psi_{1\ 10},\phi_{1\ 10}$ 40	0	40

Notes   (a) $\bar{r}_j$ is prescribed.

(b) It should be considered whether there are solutions, though the parameters are being chosen arbitrarily.

Table 1.3   Motion Generation of Inverted Two-Gear Five-Bar Linkage

No. of precision positions	No. of algebraic equations	Unknowns and their numbers	Arbitrarily chosen reals and their numbers	No. of unknowns to be solved for
ϕ_{1j}, ψ_{1j}, n prescribed; linear solution				
1	4	$\bar{L}_0, \bar{L}_1, \bar{L}_{21}, \bar{L}_{31}$ $, \bar{L}_{41}, \bar{L}_{51}$ 12	4 vectors of the left 8	4
2	8	Above 12	2 vectors of the left 4	8
3	12	Above 12	0	12
ϕ_{1j} prescribed, non-linear solution				
4	16	Above $+ \psi_{12}, \psi_{13}, \psi_{14},$ n 16	0	16
Nothing prescribed, non-linear solution				
5	20	Above $+ \psi_{15}, \phi_{12}, \phi_{13},$ ϕ_{14}, ϕ_{15} 21	n 1	20

Notes   (a) $\bar{r}_j$ is prescribed. In addition, γ_{1j} is also prescribed for controlling rigid body's movement.

(b) It should be considered whether there are solutions, though the parameters are being chosen arbitrarily

.

1.1.2.2   Function Generators

Being a function generator, the four-bar linkage is the simplest. Studying it first would be a nice beginning. Let the vector equation of closure of the four-bar linkage is

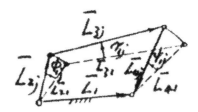

Figure 1.12 Four-bar linkage function generator

$$\overline{L}_{2j} + \overline{L}_{3j} + \overline{L}_{4j} = \overline{L}_{1j} = \overline{L}_1 = 1 \qquad (1.10)$$

where

$\overline{L}_1$ can be set equal to unity , just for easy calculation. The equation can also be written as

$$e^{i\phi_{1j}}\overline{L}_{21} + e^{i\gamma_{1j}}\overline{L}_{31} + e^{i\psi_{1j}}\overline{L}_{41} = 1 \qquad (1.11)$$

and for the first position, we have

$$\overline{L}_{21} + \overline{L}_{31} + \overline{L}_{41} = 1$$

The original function $y = f(x)$ is to be prescribed of course. Once the ranges of x ,y are being prescribed as $\Delta x, \Delta y$, and that of ϕ, ψ are being chosen as $\Delta\phi, \Delta\psi$; then

the scale factors $R_\phi (= \dfrac{\Delta\phi}{\Delta x}), R_\psi (= \dfrac{\Delta\psi}{\Delta y})$ can be defined and the conversed function

$\psi = f(\phi)$ is obtained . Then ϕ_{1j} and ψ_{1j} will be determined by the selected

precision positions, i.e. precision points. Hence, the decision parameters R_ϕ and R_ψ

would replace ϕ_{1j} and ψ_{1j} in discussing. The following table gives the relationships

of all the parameters involved.

Table 1.4   Planar Four-Bar Linkage Function Generator

No. of precision positions	No. of algebraic equations	Unknowns and their numbers	Arbitrarily chosen reals and their number	No. of unknowns to be solved for

R_ϕ, R_ψ prescribed; linear solution				
1	2	$\overline{L}_{21}, \overline{L}_{31}, \overline{L}_{41}$ 6	2 vectors of left, 4	2
2	4	Above $+ \gamma_{12}$ 7	1 vector of left $+ \gamma_{12}$ 3	4
3	6	Above $+ \gamma_{13}$ 8	γ_{12}, γ_{13} 2	6
R_ϕ, R_ψ prescribed; non-linear solution				
4	8	Above $+ \gamma_{14}$ 9	γ_{12} 1	8
5	10	Above $+ \gamma_{15}$ 10	0	10
Nothing prescribed, non linear solution				
6	12	Above $+ \gamma_{16}, R_\phi, R_\psi$ 13	γ_{12} 1	12
7	14	Above $+ \gamma_{17}$ 14	0	14

Now the mechanism's three finitely separated precision positions synthesized equations of first order approximation appears here in matrix.

$$\begin{bmatrix} 1 & 1 & 1 \\ e^{j\phi_{12}} & e^{i\gamma_{12}} & e^{i\psi_{12}} \\ e^{i\phi_{13}} & e^{i\gamma_{13}} & e^{i\psi_{13}} \end{bmatrix} \begin{bmatrix} \overline{L}_{21} \\ \overline{L}_{31} \\ \overline{L}_{41} \end{bmatrix} = \begin{bmatrix} 1 \\ 1 \\ 1 \end{bmatrix} \qquad (1.12)$$

Once the number of the precision positions of the four-bar linkage function generator is over three, the synthesis on the mechanism will be tedious because some compatible non-linear equations should be added for problem solving.

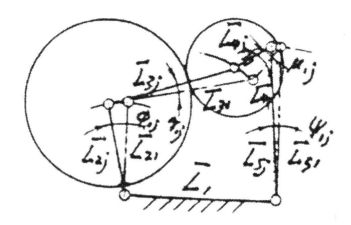

Figure 1.13 Inverted two-gear five-bar function generator

However, the number of precision positions even up to four is available for the inverted two-gear five-bar linkage having linear solutions. Of course, it is welcome. See the following:

Table 1.5   Inverted Two-Gear Five-Bar Function Generator

No .of precision positions	No. of algebraic equations	Unknowns and their numbers	Arbitrarily chosen reals and their numbers	No. of unknowns to be solved for
R_ϕ , R_ψ prescribed; linear solution				
1	2	$\bar{L}_{21}, \bar{L}_{31}, \bar{L}_{41}, \bar{L}_{51}$, n 9	3 vectors of the left, n 7	2
2	4	Above $+\gamma_{12}$, 10	2 vectors of the left, n, γ_{12} 6	4
3	6	Above $+\gamma_{13}$ 11	1 vector of the left, $n, \gamma_{12}, \gamma_{13}$, 5	6
4	8	Above $+\gamma_{14}$ 12	$n, \gamma_{12}, \gamma_{13}, \gamma_{14}$ 4	8
R_ϕ , R_ψ prescribed; non-linear solution				
5	10	Above $+\gamma_{15}$ 13	$n, \gamma_{12}, \gamma_{13}$ 3	10

6	12	Above $+\gamma_{16}$ 14	n,γ_{12}	2	12
7	14	Above $+\gamma_{17}$ 15	n	1	14
8	16	Above $+\gamma_{18}$ 16		0	16
Nothing prescribed; non-linear solution					
9	18	Above $+R_{\phi},R_{\psi},\gamma_{19}$ 19	n	1	18
10	20	Above $+\gamma_{1\,10}$ 20		0	20

The four finitely separated precision position synthesized equations of first-order approximation   are presented in matrix here

$$
\begin{bmatrix}
1 & 1 & 1 & 1 \\
e^{i\phi_{12}} & e^{i\gamma_{12}} & e^{i\mu_{12}} & e^{i\psi_{12}} \\
e^{i\phi_{13}} & e^{i\gamma_{13}} & e^{i\mu_{13}} & e^{i\psi_{13}} \\
e^{i\phi_{14}} & e^{i\gamma_{14}} & e^{i\mu_{14}} & e^{i\psi_{14}}
\end{bmatrix}
\begin{bmatrix}
\overline{L}_{21} \\
\overline{L}_{31} \\
\overline{L}_{41} \\
\overline{L}_{51}
\end{bmatrix}
=
\begin{bmatrix}
\overline{L}_{11} \\
\overline{L}_{11} \\
\overline{L}_{11} \\
\overline{L}_{11}
\end{bmatrix}
\tag{1.13}
$$

where $\overline{L}_{11}=\overline{L}_{1}=1$

【Example 1.1】 Design an inverted two-gear five-bar function generator to generate $y=x^2$, for $0\le x\le 1$, with four precision positions.

Solution: Calculate the spacing of the precision positions with Chebyshev spacing equation.

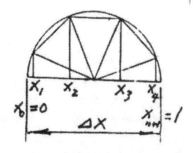

Figure 1.14   Four precision positions with Chebyshev spacing

As
$$x_j = \frac{1}{2}(x_0 + x_{n+1}) - \frac{1}{2}(x_{n+1} - x_0)\cos(\frac{2j-1}{2n})\pi$$

Where

$j=1,2\ldots\ldots n, n$ is 4 here, $\quad x_0 = 0, x_{n+1} = 1$

Then

$$x_1 = \frac{1}{2}(0+1) - \frac{1}{2}(1-0)\cos(\frac{2\times 1 - 1}{2\times 4})\pi = 0.5 - 0.5\cos(\frac{1}{8})\pi = 0.0380602$$

$$x_2 = 0.5 - 0.5\cos(\frac{3}{8})\pi = 0.3086582$$

$$x_3 = 0.5 - 0.5\cos(\frac{5}{8})\pi = 0.6413417$$

$$x_4 = 0.5 - 0.5\cos(\frac{7}{8})\pi = 0.9619397$$

whereas

$$y_1 = x_1^2 = (0.0380602)^2 = 0.00144857$$
$$y_2 = (0.3086582)^2 = 0.0952698$$
$$y_3 = (0.6913417)^2 = 0.4779533$$
$$y_4 = (0.9619397)^2 = 0.9253279$$

Both the ranges of x and y, Δx and Δy, are equal to 1; and ranges those of input

and output angles, $\Delta\phi$ and $\Delta\psi$, 90^0. Then from

$$\phi_{1j} = \frac{\Delta\phi}{\Delta x}(x_j - x_1) = 90^0(x_j - x_1)$$
$$\psi_{1j} = \frac{\Delta\psi}{\Delta y}(y_j - y_1) = 90^0(y_j - y_1)$$

We have

$$\phi_{12} = 90^0(0.3086582 - 0.0380602) = 24.35382^0$$
$$\phi_{13} = 58.795335^0$$
$$\phi_{14} = 83.149155^0$$
$$\psi_{12} = 8.443917^0$$
$$\psi_{13} = 42.885432^0$$
$$\psi_{14} = 83.149146^0$$

Let

$$\gamma_{12} = 20^0, \gamma_{13} = 40^0, \gamma_{14} = 60^0, n = -1$$

for

$$\mu_{1j} = \gamma_{1j} + (\phi_{1j} - \gamma_{1j})n$$

we have

$$\mu_{12} = 20^0 + (24.35382^0 - 20^0)(-1) = 15.64018^0$$

$$\mu_{13} = 40^0 + (58.795335^0 - 40^0)(-1) = 21.204665^0$$

$$\mu_{14} = 60^0 + (83.149155^0 - 60^0)(-1) = 36.850845^0$$

Now by means of Appendix C, rewrite eq.(1.13) as eight linear algebraic equations for solving, we have

$L_{1x} = 1.0$ $L_{1y} = 0.0$

$L_{21x} = -32.742382$ $L_{21y} = 23.813286$

$L_{31x} = 67.8865356$ $L_{31y} = -39.3288918$

$L_{41x} = -36.3696175$ $L_{41y} = 12.1787043$

$L_{51x} = 2.2254632$ $L_{51y} = 3.3369584$

Considering these solutions, seeing that the ratio between link 3 and frame 1 is too big. If n equals $-\frac{1}{2}, -1, -2$; $\gamma_{12}, \gamma_{13}, \gamma_{14}$ equals 0, $\pm 20^0, \pm 40^0$ separately; then a series of solution spectra are to be obtained, the results are some better, some worse, and some even no solution. It is likely that γ might be too big. Let us try to

set $\gamma_{12} = 0^0, \gamma_{13} = 10^0, \gamma_{14} = 15^0, n = -1$, then

$$\mu_{12} = -24.35382^0$$

$$\mu_{13} = -38.79533^0$$

$$\mu_{14} = -53.149155^0$$

$L_{1x} = 1.0$ $L_{1y} = 0.0$

$L_{21x} = 0.6658918$ $L_{21y} = 0.5896330$

$L_{31x} = 0.0914043$ $L_{31y} = -0.6491353$

$L_{41x} = 0.3739460$ $L_{41y} = 0.6133349$

$L_{51x} = -0.1317421$ $L_{51y} = -0.5538324$

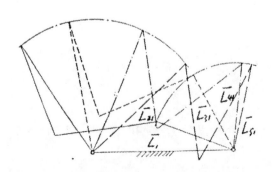

Figure 1.15   Inverted two-gear five-bar linkage function generator, generating function $y = x^2$, interval $0 \le x \le 1, \gamma_{12} = 0^0, \gamma_{13} = 10^0, \gamma_{14} = 15^0, n = -1$, Shown in its first(long dashed); second (dot dashed), third (short dashed), and fourth (light weight) precision positions. Gears not shown.

The errors on X and Y axes are 0.048% and 0.933% respectively.

1.1.2.3   Synthesis on Velocity, Acceleration and Higher- Order

Until now, synthesis is to aim at the precision positions; that is, moving particle's displacement (definite separations). If not only consider the precision point (i.e. position), but the moving point's (i.e. particle's) velocity, acceleration and jerk also, then synthesis will come to higher order. Velocity is the first derivative of displacement to time, acceleration is the second, and the jerk is the third.

As we know that the number of precision positions of function generator synthesis is only a few. If velocity and/or acceleration requirement is needed, the number of precision positions must be reduced. Let the number of precision positions be four, then P-P-P-P is represented for the evenly distributed state. When velocity, acceleration, and even higher orders of synthesis are involved, then PP, PPP, and PPPP represent 2, 3, and 4 infinitesimally approaching positions respectively. Fig.1.16 shows the five cases of multiply separated position, containing finitely and infinitesimally separated position, and their associated error curves. In the figure, solid line is the theoretical value, short dashed line is the actual value, and ε is the difference between them. Besides, there are errors due to manufacturing and assembling, as so called "mechanical error", will not be discussed in this text.

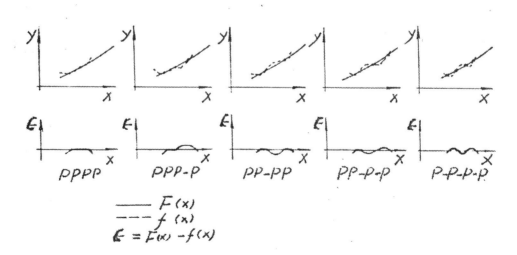

Figure 1.16   Cases of multiply separated positions and their associated error curves

Now we go on discussing the multiply separated precision positions of the inverted two-gear five-bar linkage. Velocity has to be shown in vector equations if the velocity synthesis of one position (say, the second position in the following example) is required. Owing to the restriction of linear solution, subtract one from the four in eq.(1.13). The new synthesis equations containing the velocity vector in matrix will be

$$
\begin{bmatrix}
1 & 1 & 1 & 1 \\
e^{i\phi_{12}} & e^{i\gamma_{12}} & e^{i\mu_{12}} & e^{i\psi_{12}} \\
e^{i\phi_{13}} & e^{i\gamma_{13}} & e^{i\mu_{13}} & e^{i\psi_{13}} \\
e^{i\phi_{12}} & \dot{\gamma}_2^* e^{i\gamma_{12}} & \dot{\mu}_2^* e^{i\mu_{12}} & \dot{\psi}_2^* e^{i\psi_{12}}
\end{bmatrix}
\begin{bmatrix}
\overline{L}_{21} \\
\overline{L}_{31} \\
\overline{L}_{41} \\
\overline{L}_{51}
\end{bmatrix}
=
\begin{bmatrix}
\overline{L}_{11} \\
\overline{L}_{11} \\
\overline{L}_{11} \\
0
\end{bmatrix}
\qquad (1.14)
$$

where $\overline{L}_{11}$ is defined as 1; $\phi_{12}, \phi_{13}, \psi_{12}, \psi_{13}$, and $\dot{\phi}_2$ are prescribed; $n, \gamma_{12}, \gamma_{13}, \dot{\gamma}_2$ are to be chosen arbitrarily . Then μ_{12}, μ_{13}, and $\dot{\mu}_2$ can be calculated by

equations $\qquad \mu_{1j} = \gamma_{1j} + (\phi_{1j} - \gamma_{1j})n \qquad$ and $\qquad \dot{\mu}_j = \dot{\gamma}_j + (\dot{\phi}_j - \dot{\gamma}_j)n$

whereas $\dot{\psi}_2$ is obtained by the method shown in the following example.

---

*Written as $\dot{\gamma}_2, \dot{\mu}_2, \dot{\psi}_2$ due to the fact that the velocities are in position 2

[Example 1.2] Design an inverted two-gear five-bar function generator to generate $y = x^2$ for $0 \le x \le 1$, with three precision positions and a velocity approximation at one of them (say, second)

Solution: From Chebyshev spacing equation

$$x_j = \frac{1}{2}(x_0 + x_{n+1}) - \frac{1}{2}(x_{n+1} - x_0)\cos(\frac{2j-1}{2n})\pi$$

where $\ j = 1,2,3, n = 3, x_0 = 0, x_{n+1} = 1$

we have

$$x_1 = 0.0669872$$
$$x_2 = 0.5$$
$$x_3 = 0.9330127$$

And from $\ y = x^2, \dot{y} = 2x$, we obtain

$$y_1 = 0.0\,0\,4\,4\,8\,7$$
$$y_2 = 0.2\,5$$
$$y_3 = 0.8\,7\,0\,5\,1\,2$$
$$\dot{y}_1 = 0.1\,3\,3\,9\,7\,4$$
$$\dot{y}_2 = 1.0$$
$$\dot{y}_3 = 1.8\,6\,6\,0\,2\,5$$

Furthermore, from

$$\phi_{1j} = \frac{\Delta\phi}{\Delta x}(x_j - x_1)$$

$$\psi_{1j} = \frac{\Delta\psi}{\Delta y}(y_j - y_1)$$

let $\qquad \Delta\phi = 90^0, \Delta\psi = 90^0$

whereas $\qquad \Delta x = 1, \Delta y = 1$

then

$$\phi_{1j} = 90^0(x_j - x_1)$$

$$\psi_{1j} = 90^0(y_j - y_1)$$

$$\dot{\psi} = \frac{d\psi}{d\phi} = (\frac{d\psi}{dy})(\frac{dy}{dx})(\frac{dx}{d\phi})$$

Since the relationships between x and ϕ, y and ψ, both are linear.

$$\frac{d\psi}{dy} = \frac{90^0}{1} ; \frac{dx}{d\phi} = \frac{1}{90^0}$$

hence $\qquad \dfrac{d\psi}{d\phi} = \dfrac{dy}{dx}$

Now we can get all the parameters

$$n = -1$$

$$\dot{\phi}_2 = 1$$

$$\gamma_{12} = 15^0, \gamma_{13} = 30^0, \dot{\gamma}_2 = 0.5 \qquad \text{(arbitrarily chosen)}$$

$$\phi_{12} = 90^0(0.5 - 0.0669872) = 38.971152^0$$
$$\phi_{13} = 90^0(0.9330127 - 0.0669872) = 77.950345^0$$

$$\mu_{12} = \gamma_{12} + (\phi_{12} - \gamma_{12})n = 15^0 + (38.974115^0 - 15^0)(-1) = -8.979115^0$$
$$\mu_{13} = 30^0 + (77.950345^0 - 30^0)(-1) = 17.950393^0$$

$$\dot{\mu}_2 = \dot{\gamma}_2 + (\dot{\phi}_2 - \dot{\gamma}_2)n = 0.5 + (1 - 0.5)(-1) = 0$$

$$\psi_{12} = 90^0(0.25 - 0.00448728) = 22.096952^0$$

$$\psi_{13} = 90^0(0.8705127 - 0.00448728) = 77.9422950$$

$$\dot{\psi}_2 = \dot{y}_2 = 1$$

Then by means of Appendix C, rewrite eq.(1.14) as eight linear algebraic equations. Solving them, we get

$$L_{21x} = 2.1713457 \qquad\qquad L_{21y} = -0.1811433$$
$$L_{31x} = -3.1301391 \qquad\qquad L_{31y} = 0.8321379$$
$$L_{41x} = 2.5878487 \qquad\qquad L_{41y} = 0.4096750$$
$$L_{51x} = 0.6290553 \qquad\qquad L_{51y} = -1.0656793$$

L_1 is still equal to unity. Errors on X and Y axes are 0.0002% and 0.00101% respectively.

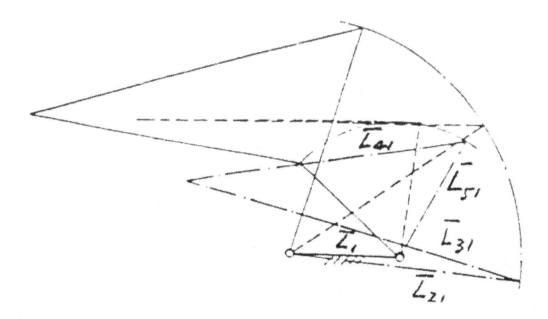

Figure 1.17 Three precision positions of the inverted two-gear five-bar function generator, generating $y = x^2, 0 \le x \le 1, n = -1, \gamma_{12} = 15^0, \gamma_{13} = 30^0, \dot{\gamma}_2 = 0.5$. Gears not shown. First (dot dashed), second (short dashed), third (light weight) positions are shown.

If the demand of certain precision position is not only for velocity, but also for acceleration, then the number of precision positions will be reduced of one more. The synthesis equations in matrix become

$$\begin{bmatrix} 1 & 1 & 1 & 1 \\ 1 & \dot{\gamma}_1 & \dot{\mu}_1 & \dot{\psi}_1 \\ -1 & -(\dot{\gamma}_1)^2 + i\ddot{\gamma}_1 & -(\dot{\mu}_1)^2 + i\ddot{\mu}_1 & -(\dot{\psi}_1)^2 + i\ddot{\psi}_1 \\ e^{i\phi_{12}} & e^{i\gamma_{12}} & e^{i\mu_{12}} & e^{i\psi_{12}} \end{bmatrix} \begin{bmatrix} \overline{L}_{21} \\ \overline{L}_{31} \\ \overline{L}_{41} \\ \overline{L}_{51} \end{bmatrix} = \begin{bmatrix} \overline{L}_1 \\ 0 \\ 0 \\ \overline{L}_1 \end{bmatrix}$$

(1.15)

if the original four precision positions are used; where $\overline{L}_1$ is still the frame length

unit 1. And n prescribed; $\gamma_{12}, \dot{\gamma}_1, \ddot{\gamma}_1$ arbitrarily chosen; $\phi_{12}, \mu_{12}, \psi_{12}, \dot{\mu}_1, \ddot{\mu}_1, \dot{\psi}_1, \ddot{\psi}_1$

calculated from $y = f(x)$ and $\mu = f(\phi, \gamma)$; substituting all the parameters into eqs.

of (1.15), the lengths of links will be obtained.

1.1.3 Synthesized Non-Linear Equations of Cycloidal Crank Two-Gear Five-Bar
Linkage

As we have stated above, if the number of precision positions of the inverted two-gear
five-bar linkage is greater than four, the tedious nonlinear equations will occur. We
can see the difficulty of this kind in Appendix D. There are many scholars, however,
are still interested in searching for more precision positions synthesis.

While in this section a method of *extension of Freudenstein's equation* is being
introduced, whereas the cycloidal crank two-gear five-bar linkage plays an important
role because it is popular and utilized more for its good steadiness behavior and
cycloidal simplicity.

1.1.3.1 The Displacement Equation for Cycloidal Crank Two-Gear Five-Bar Function
Generator
——Extended Freudenstein's Equation

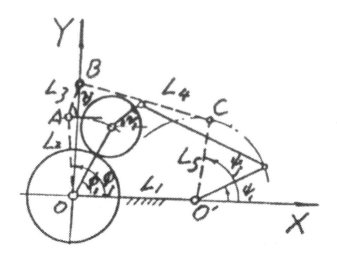

Figure 1.18   Cycloidal crank two-gear five-bar linkage in its first
(solid) and jth (dashed) precision positions

The displacement equation for cycloidal crank two-gear five-bar linkage (shown in Fig.1.18) is expressed by the moving joints A,B,C relative to the static coordinate system XOY as follows:

For A,B and C, we have

$$x_A = L_2 \cos\phi_j \qquad\qquad y_A = L_2 \sin\phi_j$$
$$x_B = L_2 \cos\phi_j + L_3 \cos\gamma_j \qquad\qquad y_B = L_2 \sin\phi_j + L_3 \sin\gamma_j$$
$$x_C = L_1 + L_5 \cos\psi_j \qquad\qquad y_C = L_5 \sin\psi_j$$

Assume link 4 is an ideal rigid body, and its length is constant, so

$$(x_C - x_B)^2 + (y_C - y_B)^2 = L_4^2$$

substituting for x_B, x_C, y_B and y_C, leads to

$$(L_1 + L_5 \cos\psi_j - L_2 \cos\phi_j - L_3 \cos\gamma_j)^2 + (L_5 \sin\psi_j - L_2 \sin\phi_j - L_3 \sin\gamma_j)^2 = L_4^2$$

Simplified and the trigonometric identities being used, we have

$$D\sin\psi_j + E\cos\psi_j = F \qquad\qquad\qquad (1.16)$$

where

$$D = \sin\phi_j + \frac{L_3}{L_2}\sin\gamma_j$$

$$E = -\frac{L_1}{L_2} + \cos\phi_j + \frac{L_3\cos\gamma_j}{L_2}$$

$$F = \frac{L_3}{L_5}\cos(\phi_j - \gamma_j) - \frac{L_1}{L_5}\cos\phi_j - \frac{L_1 L_3}{L_2 L_5}\cos\gamma_j + \frac{L_1^2 + L_2^2 + L_3^2 - L_4^2 + L_5^2}{2L_2 L_5}$$

whereas $\quad \sin\psi_j = \dfrac{2\tan\dfrac{\psi_j}{2}}{1+\tan^2(\dfrac{\psi_j}{2})} \quad , \quad \cos\psi_j = \dfrac{1-\tan^2(\dfrac{\psi_j}{2})}{1+\tan^2(\dfrac{\psi_j}{2})}$

substituting these into eq.(1.16), we have

$$\tan\frac{\psi_j}{2} = \frac{D \pm (D^2 + E^2 - F^2)^{\frac{1}{2}}}{E + F}$$

Obviously, ψ_j has two values. These two represent two configurations. Note that D,

E, and F contain the input angle, link lengths, and the gear ratio (i.e. $\dfrac{N_1}{N_3}$ in

$$\gamma_{1j} (= \phi_{1j}(1 + \frac{N_1}{N_3}))$$

If we want to obtain link lengths for angular positions of input crank , cyclodial crank, and output crank angles (i.e. ϕ_j, γ_j and ψ_j) respectively in the jth position of the mechanism; let us rearrange eq. (1.16) as follows:

$$\frac{L_3}{L_5}\cos(\phi_j - \gamma_j) - \frac{L_3}{L_2}\cos(\gamma_j - \psi_j) + \frac{L_1}{L_2}\cos\psi_j - \frac{L_1}{L_5}\cos\phi_j - \frac{L_1 L_3}{L_2 L_5}\cos\gamma_j$$

$$-\cos(\phi_j - \psi_j) + \frac{L_1^2 + L_2^2 + L_3^2 - L_4^2 + L_5^2}{2 L_2 L_5} = 0$$

$$(1.17)$$

where ϕ_j, γ_j and ψ_j are the position angles of input crank, cycloidal crank and output crank respectively in jth position of the mechanism. Note that this equation can be reduced to Freudenstein's equation (Appendix E) for L_3 equals zero, i.e. the displacement equation for four-bar linkage.

The more useful form of eq (1.17) is the following expression:

$$K_1 \cos(\phi_j - \gamma_j) - K_2 \cos(\gamma_j - \psi_j) + K_3 \cos\psi_j + K_4$$
$$= \cos(\phi_j - \psi_j) + K_5 \cos\gamma_j + K_6 \cos\phi_j$$

$$(1.18)$$

where

$$K_1 = \frac{L_3}{L_5}$$

$$K_2 = \frac{L_3}{L_2}$$

$$K_3 = \frac{L_1}{L_2}$$

$$(1.19\text{-}1)$$

$$K_4 = \frac{L_1^2 + L_2^2 + L_3^2 - L_4^2 + L_5^2}{2 L_2 L_5}$$

$$K_5 = \frac{L_1 L_3}{L_2 L_5}$$

$$(1.19\text{-}2)$$

$$K_6 = \frac{L_1}{L_5}$$

Eq(1.18) is the displacement equation for cycloidal crank two-gear five-bar linkages.

Furthermore, we can extent it to three-gear, four-gear associate to four-bar, six-bar combinations if possible.

1.1.3.2 Four-Precision-Position Approximation of Cycloidal Crank Two-Gear Five-Bar Function Generator

Though the four-position approximation (i.e. four prescribed position synthesis) of this mechanism can be managed by the method other than the extended Freudenstein's equation, using it may give some good ideas for comparison. With prescribed input crank angles ϕ_1, ϕ_2, ϕ_3, and ϕ_4, output angles ψ_1, ψ_2, ψ_3, and ψ_4, gear ratio N_1/N_3, first cycloidal crank angle γ_1, and the calculated γ_j; equations containing six unknowns ($K_1, K_2, \ldots K_6$) can be written as follows:

$$K_1 \cos(\phi_j - \gamma_j) - K_2 \cos(\gamma_j - \psi_j) + K_3 \cos\psi_j + K_4$$
$$= \cos(\phi_j - \psi_j) + K_5 \cos\gamma_j + K_6 \cos\phi_j$$

$$(1.20)$$

where j=1, 2, 3, 4

It can not be solved unless the following two compatible equations are added.

$$K_5 = K_1 K_3$$

$$K_1 K_3 - K_5 = 0 \qquad (1.21)$$

$$K_6 = \frac{K_1 K_3}{K_2},$$

$$K_2 K_6 - K_1 K_3 = 0 \qquad (1.22)$$

Then all these six equations can be solved for six unknowns.

Let $\quad \lambda_1 \equiv K_5, \qquad \lambda_2 \equiv K_6$

the compatible equations become

$$K_1 K_3 - \lambda_1 = 0 \qquad (1.23)$$

$$K_2 \lambda_2 - K_1 K_3 = 0$$

$$(1.24)$$

Now eq (1.20) can be written as

$$K_1 \cos(\phi_j - \gamma_j) - K_2 \cos(\gamma_j - \psi_j) + K_3 \cos\psi_j + K_4 = \lambda_1 \cos\gamma_j + \lambda_2 \cos\phi_j + \cos(\phi_j - \psi_j)$$

$$(1.25)$$

where $j=1, 2, 3, 4$

According to the principle of superposition, taking each term on the right side of eq.(1.25) equals the sum of all the modified terms on the left; as there are three terms on the right, then we have three sets linear equations as follows:

$$\lambda_1[l_1\cos(\phi_j - \gamma_j) - l_2\cos(\gamma_j - \psi_j) + l_3\cos\psi_j + l_4] = \lambda_1\cos\gamma_j$$

$$\lambda_1[l_1\cos(\phi_j - \gamma_j) - l_2\cos(\gamma_j - \psi_j) + l_3\cos\psi_j + l_4] =_1\cos\gamma_j \tag{1.26}$$

$$\lambda_2[m_1\cos(\phi_j - \gamma_j) - m_2\cos(\gamma_j - \psi_j) + m_3\cos\psi_j + m_4] = \lambda_2\cos\phi_j$$

$$m_1\cos(\phi_j - \gamma_j) - m_2\cos(\gamma_j - \psi_j) + m_3\cos\psi_j + m_4 = \cos\phi_j \tag{1.27}$$

$$n_1\cos(\phi_j - \gamma_j) - n_2\cos(\gamma_j - \psi_j) + n_3\cos\psi_j + n_4 = \cos(\phi_j - \psi_j) \tag{1.28}$$

where $j=1, 2, 3, 4$.

There will be twelve equations for the three sets. Twelve unknowns l_p, m_p, n_p ($p=1$, 2, 3, 4) can be solved, whereas $K_p = \lambda_1 l_p + \lambda_2 m_p + n_p$

$$(1.29)$$

Substituting K_p into eqs. (1.23) and (1.24), we have

$$(\lambda_1 l_1 + \lambda_2 m_1 + n_1)(\lambda_1 l_3 + \lambda_2 m_3 + n_3) - \lambda_1 = 0$$

$$(1.30)$$

$$\lambda_2(\lambda_1 l_2 + \lambda_2 m_2 + n_2) - (\lambda_1 l_3 + \lambda_2 m_3 + n_3)(\lambda_1 l_1 + \lambda_2 m_1 + n_1) = 0 \tag{1.31}$$

Let λ_2 be the unknown, rewrite the above two equations in polynomials as follows:

$$m_1 m_3 \lambda_2^2 + [\lambda_1(m_1 l_3 + l_1 m_3) + n_3 m_1 + m_3 n_1]\lambda_2 + [\lambda_1^2 l_1 l_3 + \lambda_1(l_3 n_1 + n_3 l_1 - 1) + n_1 n_3] = 0$$

$$(1.32)$$

$$(m_2 - m_1 m_2)\lambda_2^2 + [\lambda_1(l_2 - l_1 m_3 - m_1 l_3) + n_2 - m_1 n_3 - n_1 m_3]\lambda_2 - [\lambda_1^2 l_3 l_1 + \lambda_1(l_1 n_3 + n_1 l_3) + n_3 n_1] = 0$$

$$(1.33)$$

Using sylvester's dialytic elimination, see Appendix F, we get

$$\begin{vmatrix} 0 & m_1m_3 & \left[\begin{array}{l}(m_1l_3+l_1m_3)\lambda_1\\+n_3m_1+m_3n_1\end{array}\right] & \left[l_1l_3\lambda_1^2+\left(\begin{array}{l}l_3n_1\\+n_3l_1-1\end{array}\right)\lambda_1+n_1n_3\right] \\ 0 & m_2-m_1m_3 & \left[\begin{array}{l}(l_2-l_1m_3-m_1l_3)\lambda_1\\+n_2-m_1n_3-n_1m_3\end{array}\right] & -\left[\begin{array}{l}l_1l_3\lambda_1^2+\left(\begin{array}{l}l_1n_3\\+n_1l_3\end{array}\right)\lambda_1\\+n_1n_3\end{array}\right] \\ m_1m_3 & \left[\begin{array}{l}(m_1l_3+l_1m_3)\lambda_1\\+n_3m_1+m_3n_1\end{array}\right] & \left[l_1l_3\lambda_1^2+\left(\begin{array}{l}l_3n_1\\+n_3l_1\\-1\end{array}\right)\lambda_1+n_1n_3\right] & 0 \\ m_2-m_1m_3 & \left[\begin{array}{l}(l_2-l_1m_3-m_1l_3)\lambda_1\\+n_2-m_1n_3-n_1m_3\end{array}\right] & -\left[\begin{array}{l}l_1l_3\lambda_1^2+\left(\begin{array}{l}l_1n_3\\+n_1l_3\end{array}\right)\lambda_1\\+n_1n_3\end{array}\right] & 0 \end{vmatrix} =0 \qquad (1.34)$$

Expanding it, then a quartic polynomial of λ_1 is shown.

$$R_1\lambda_1^4 + R_2\lambda_1^3 + R_3\lambda_1^2 + R_4\lambda_1 + R_5 = 0$$

$$(1.35)$$

This fourth degree polynomial in λ_1 may have λ_1's real roots of four, or two, or even none at all. Two roots of λ_2 may be presented in either of the eqs. (1.32) and (1.33) when each real root of λ_1 is substituted into. And one of the two real roots of λ_2 can fulfill both eqs. (1.32) and (1.33). By substituting pairs of λ_1 and λ_2 into eq.(1.29), K_p is to be found. Finally, the link lengths can be calculated by eq (1.19) as K_p is known.

An example is to be prepared for a good understanding. And it is necessary to have a look at a simple one, Appendix D, beforehand .

【Example 1.3】 Design a cycloidal crank two-gear five-bar function generator to generate $y = x^2$, $0 \le x \le 1$, four precision positions.

Solution

Let $\phi_1 = 30^0, \gamma_1 = 60^0, \psi_1 = 30^0, \Delta\phi = 90^0, \Delta\psi = 90^0$.

Same as example 1.1, we have

$$x_1 = 0.0380602 \qquad y_1 = 0.00144857$$
$$x_2 = 0.3086582 \qquad y_2 = 0.0952648$$
$$x_3 = 0.6913417 \qquad y_3 = 0.4779533$$
$$x_4 = 0.9619397 \qquad y_4 = 0.9253279$$

Again, according to example 1.1 we have ϕ_{1j} and ψ_{1j}.

Whereas $\phi_j = \phi_1 + \phi_{1j}, \psi_j = \psi_1 + \psi_{1j}, \gamma_j = \gamma_1 + \gamma_{1j}, \gamma_{1j} = 2\phi_{1j}$

(for $\gamma_{1j} = \phi_{1j}(1 + \dfrac{N_1}{N_3}) = \phi_{1j}(1+1) = 2\phi_{1j}$, where $N_1 = N_3$), we have

$$\phi_{12} = 24.35382^0 \qquad \phi_2 = 30^0 + 24.35382^0 = 54.35382^0$$
$$\phi_{13} = 58.795335^0 \qquad \phi_3 = 88.795335^0$$
$$\phi_{14} = 83.149155^0 \qquad \phi_4 = 113.149155^0$$
$$\psi_{12} = 8.443917^0 \qquad \psi_2 = 30^0 + 8.443917^0 = 38.443917^0$$
$$\psi_{13} = 42.885432^0 \qquad \psi_3 = 72.885432^0$$
$$\psi_{14} = 83.149146^0 \qquad \psi_4 = 113.149146^0$$
$$\gamma_{12} = 2\phi_{12} = 48.70764^0 \qquad \gamma_2 = 60^0 + 48.70764^0 = 108.70764^0$$
$$\gamma_{13} = 117.59067^0 \qquad \gamma_3 = 177.59067^0$$
$$\gamma_{14} = 166.29831^0 \qquad \gamma_4 = 226.29831^0$$

By eq.(1.26), we get four linear equations as follows:

$$l_1 \cos(30^0 - 60^0) l_2 \cos(60^0 - 30^0) + l_3 \cos 30^0 + l_4 = \cos 60^0$$

$$l_1 \cos(54.35382^0 - 108.70764^0) - l_2 \cos(108.70764^0 - 38.443917^0)$$
$$+ l_3 \cos 38.443917^0 + l_4 = \cos 108.70764^0$$

$$l_1 \cos(88.795335^0 - 177.59067^0) - l_2 \cos(177.59067^0 - 72.885432^0)$$
$$+ l_3 \cos 72.885432^0 + l_4 = \cos 177.549066^0$$

$$l_1 \cos(113.149155^0 - 226.29831^0) - l_2 \cos(226.29831^0 - 113.149155^0)$$
$$+ l_3 \cos 113.149155^0 + l_4 = \cos 226.29831^0$$

Solving these, we have

$$l_1 = 1.686674$$
$$l_2 = -0.9075666$$
$$l_3 = -1.648442$$
$$l_4 = -0.3190851$$

Again, by eqs.(1.27) and (1.28), four linear equations each are obtained. Then the results

$$m_1 = 0.9999999 \qquad\qquad n_1 = -0.3309020$$
$$m_2 = 0.0 \qquad\qquad n_2 = 0.2348504$$
$$m_3 = 0.0000000596 \qquad\qquad n_3 = 0.0960525$$
$$m_4 = 0.0 \qquad\qquad n_4 = 1.0$$

Substituting the above parameters into eq.(1.34) and expanding it, we have

$$-(1.3650273E-07)\lambda_1^4 + (4.16052565E-02)\lambda_1^3 - (2.540963E-10)\lambda_1^2$$
$$-(3.1137868E-09)\lambda_1 - (1.0448940E-10) = 0$$

and two real roots of λ_1

$$\left\{ \begin{array}{l} \text{-3.3758424E-02} \\ \text{-0.2587695} \end{array} \right.$$

Substituting these two values into eqs(1.32) and (1.33) separately, either has two pairs real roots of λ_2.

The one (when $\lambda_1 = -3.3758424E-02$) has

$$\left\{ \begin{array}{ll} 0.1653104, & \text{-2545127} \\ 1.1653104, & \text{-5971242} \end{array} \right.$$

Obviously, 0.1653104 is the common root of eqs.(1.32) and (1.33). So substituting them into eq.(1.29), we obtain

$$K_1 = -0.2225321$$
$$K_2 = -0.2042124$$
$$K_3 = 0.15170124$$
$$K_4 = 1.010772$$

Results: $L_1 = 1.00, L_2 = 6.591898, L_3 = -1.346147, L_4 = 1.499177, L_5 = 6.049228$

The another one (when $\lambda_1 = -0.2587695$) has

$$\begin{cases} 0.2722770, & -8768093 \\ 0.7673627, & -8768094 \end{cases}$$

With same procedure, we get the following:

$K_1 = -8768092$

$K_2 = -2.9802322E - 08$

$K_3 = -2.2351742E - 08$

$K_4 = 1.082570$

$L_1 = 1.0$

$L_2 = -3.3554428E + 07$

$L_3 = 0.9999999$

$L_4 = 3.3554428E + 07$

$L_5 = -1.1404988E - 07$

By observing answers of these two sets, we see that the former is better, the later is no good for the relative length ratio is too great. Now we change some parameter values for getting more results to compare. All these with the above are tabulated in Table 1.6.

Table 1.6   Solutions of Function Generator

$y = x^2, 0 \le x \le 1$

No	R_ϕ	R_ψ	N_1	ϕ_1	γ_1	ψ_1	L_1	L_2	L_3	L_4	L_5
1	90^0	90^0	-1	30^0	60^0	30^0	$1._0$ 1.0	$6._{591898}$ -3.3554428E+07	-1.346147 0.9999999	$1._{499173}$ 3.3554428E+07	6.049228 -1.1404988E-07
2	90^0	90^0	-1	30^0	30^0	30^0	$1._0$ 1.0	1.5992119E+07 -3.104121	$1._{429807}$ -1.405444	1.5992129E+07 1.340585	-4.073362 -4.131926
3	90^0	90^0	-1	30^0	90^0	30^0	$1._0$ 1.0	340263.1 -12.96906	0.9684302 $1._{279362}$	340259.9 1.690951	$2._{758886}$ -13.71862
4	90^0	90^0	-3	30^0	60^0	30^0	$1._0$ 1.0	-63.34206 -5.288962	1.791279 0.3932937	1264.936 0.7974325	1202.279 -5.426336
5	90^0	90^0	-3	30^0	90^0	30^0	$1._0$ 1.0	-1781.058 -5.484590	-9.352418 0.5126021	1537.040 1.037922	-249.5797 -5.662704
6	90^0	60^0	-2	30^0	60^0	30^0	$1._0$ 1.0	$89._{10430}$ -2.816266	$1._{275090}$ -1.871466	$81._{02306}$ 2.178397	$8._{327458}$ -3.168655
7	90^0	60^0	-2	30^0	60^0	30^0	$1._0$ 1.0	$8._{198127}$ -4.392969	$0._{9073936}$ -5.488808	$4._{827531}$ 5.529253	$3._{188771}$ -4.618640
8	70^0	60^0	-2	30^0	90^0	30^0	$1._0$ 1.0	-20.87593 7.178614	-2.362759 0.6656536	$2._{828967}$ 3.274660	-20.68748 3.266115
9	70^0	60^0	-2	30^0	60^0	30^0	$1._0$ 1.0	-10.03953 $6._{753811}$	-3.134391 $0._{5330519}$	$3._{266172}$ $4._{559652}$	-10.27208 $1._{705300}$

							1.0	-1.301240	0.2078747	1.116029	-1.221287
							1.0	-1.286229	0.6402192	1.141679	-2.041536
10	60°	60°	-3	30°	60°	30°	1.0	-10.72363	0.5181184	0.9406117	-11.69643
							1.0	-7.374406	0.3319734	1.314823	-6.819258
							1.0	-2.405416	0.2569000	1.087297	-2.162679
							1.0	-1.910875	0.5884352	0.8346691	-2.523995

Observing the results in Table 1.6, we select four satisfactory solutions listed below.

Table 1.7   Selected solutions

No.	Solutions belong to	Length ratios of longest to shortest
1	2nd of No. 6	3.168655
2	4th of No. 9	3.1888078
3	2nd of No. 2	4.131925
4	4th of No. 10	4.2893338

Though the lengths of link 3 to the 4th of No.9 and 4th of No.10 are both shorter than ordinary ones, they are still workable since they are rigidly attached to gear 3. Fig.1.19 *a,b,c,*, and *d* show the four best solutions in order that positions1, 2, 3, and

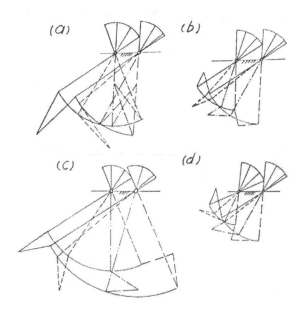

Figure 1.19   Four precision positions each of the four (a,b,c,d) cycloidal cranktwo-gear five-bar function generators generating $y = x^2$. Gears not

shown. First(solid), second (short dashed), third(dot dot long dashed),
fourth(long dashed) positions are shown.

4.In the view of harmony, b is the best. All the dials are located at the opposite sides of corresponding links just for the convenience of reading. For simplicity the range is not in full scale, i.e. from first position to 4th position only. Besides, there are some overlaps between the pairs of dials allowable if clearness is being kept.

It is hard to define which the most important factor is in so many parameters. Generally speaking, gear ratio N_1/N_3 and the initial (i.e. first) position cycloidal crank angle γ_1 may play the chief one or two. As for $\phi_1, \psi_1, \Delta\phi$, and $\Delta\psi$ may be the minors. More fully discussing and judging might be made in advanced research afterwards.

1.1.3.3 Five-Precision-Position Approximation of Cycloidal Crank Two-Gear Five-Bar Function Generator

When the precision positions reaches up to five; the ranges of input and output, $\Delta\phi$ and $\Delta\psi$, function $\psi = f(\phi)$ which is equivalent to $y = f(x)$, gear ratio n, first position input angle and cycloidal crank angle, ϕ_1 and γ_1, are all prescribed by the designer's own will. Whereas the first position output angle ψ_1 and link lengths L_2, L_3, L_4, and L_5 should be calculated by the displacement equations. The displacement equation used here is different to the above because ψ_1 is added and ψ_{1j} replace ψ_j in the new equations which contain nine unknowns in five equations as follows:

$$-K_1 \cos(\gamma_j - \psi_{1j}) + K_2 \cos\psi_{1j} + K_3 \cos(\phi_j - \gamma_j) - K_4 \sin(\phi_j - \psi_j) - K_5$$
$$= K_6 \cos\gamma_j + K_7 \cos\phi_j + K_8 \sin(\gamma_j - \psi_{1j}) + K_9 \sin\psi_{1j} + \cos(\phi_j - \psi_{1j})$$

(1.36)

j=1, 2, 3, 4, 5

where

$$K_1 = \frac{L_3}{L_2}$$

$$K_2 = \frac{L_1}{L_2}$$

$$K_3 = \frac{L_3}{L_5 \cos \psi_1}$$

$$K_4 = \tan \psi_1$$

$$K_5 = \frac{L_1^{\ 2} + L_2^{\ 2} + L_3^{\ 2} - L_4^{\ 2} + L_5^{\ 2}}{2 L_2 L_5 \cos \psi_1}$$

$$K_6 = \frac{L_1 L_3}{L_2 L_5 \cos \psi_1}$$

$$K_7 = \frac{L_1}{L_5 \cos \psi_1}$$

$$K_8 = \frac{L_3}{L_2} \tan \psi_1$$

$$K_9 = \frac{L_1}{L_2} \tan \psi_1$$

$$\left. \right\} \quad ....(1.37)$$

Unknowns K_6, K_7, K_8, K_9 have some relationships with K_1, K_2, K_3, K_4, and K_5 as follows:

$$K_6 = K_2 K_3$$

$$K_7 = \frac{K_2 K_3}{K_1}$$

$$K_8 = K_1 K_4$$

$$K_9 = K_2 K_4$$

$$\left. \right\}$$

$$(\ 1.38 \ )$$

Let $\lambda_1 \equiv K_6, \lambda_2 \equiv K_7, \lambda_3 \equiv K_8, \lambda_4 \equiv K_9$, then eq.(1.38) becomes

$$K_2 K_3 - \lambda_1 = 0$$

$$K_2 K_3 - \lambda_2 K_1 = 0$$

$$K_1 K_4 - \lambda_3 = 0$$

$$K_2 K_4 - \lambda_4 = 0$$

$$\left. \right\}$$

(1.39)

Rewrite eq.(1.36) as

$$
\left.
\begin{aligned}
&-K_1 \cos(\gamma_j - \psi_{1j}) + K_2 \cos\psi_{1j} + K_3 \cos(\phi_j - \gamma_j) - K_4 \sin(\phi_j - \psi_{1j}) + K_5 \\
&= \lambda_1 \cos\gamma_j + \lambda_2 \cos\phi_j + \lambda_3 \sin(\gamma_j - \psi_{1j}) + \lambda_4 \sin\psi_{1j} + \cos(\phi_j - \psi_{1j})
\end{aligned}
\right\}
$$

(1.40)

j=1, 2, 3, 4, 5

Following the procedure in the above section, we have five sets linear equations below.

$$
\left.
\begin{aligned}
&-l_1 \cos(\gamma_j - \psi_{1j}) + l_2 \cos\psi_{1j} + l_3 \cos(\phi_j - \gamma_j) - l_4 \sin(\phi_j - \psi_{1j}) + l_5 = \cos\gamma_j \\
&-m_1 \cos(\gamma_j - \psi_{1j}) + m_2 \cos\psi_{1j} + m_3 \cos(\phi_j - \gamma_j) - m_4 \sin(\phi_j - \psi_{1j}) + m_5 = \cos\phi_j \\
&-n_1 \cos(\gamma_j - \psi_{1j}) + n_2 \cos\psi_{1j} + n_3 \cos(\phi_j - \gamma_j) - n_4 \sin(\phi_j - \psi_{1j}) + n_5 = \sin(\gamma_j - \psi_{1j}) \\
&-p_1 \cos(\gamma_j - \psi_{1j}) + p_2 \cos\psi_{1j} + p_3 \cos(\phi_j - \gamma_j) - p_4 \sin(\phi_j - \psi_{1j}) + p_5 = \sin\psi_{1j} \\
&-q_1 \cos(\gamma_j - \psi_{1j}) + q_2 \cos\psi_{1j} + q_3 \cos(\phi_j - \gamma_j) - q_4 \sin(\phi_j - \psi_{1j}) + q_5 = \cos(\phi_j - \psi_{1j})
\end{aligned}
\right.
$$

(1.41)

j=1, 2, 3, 4, 5

Since there are twenty-five equations in these sets, then the twenty-five unknowns l_p, m_p, n_p, p_p, q_p (p=1, 2, 3, 4, 5) can be calculated. Next, obtain the following

$$
K_p = l_p \lambda_1 + m_p \lambda_2 + n_p \lambda_3 + p_p \lambda_4 + q_p
$$

(1.42)

p=1, 2, 3, 4, 5

Substituting corresponding k_p's back in eq.(1.39) , we obtain four non-linear equations containing four unknowns, $\lambda_1, \lambda_2, \lambda_3$, and λ_4. .By applying sylvester's dialytic elimination solving for these unknowns, $K_6 \sim K_9$ will be obtained by eq.(1.42). Finally, the link lengths and ψ_1 can be determined from the eqs: (1.37) and (1.38).

1.1.3.4   Commentary on Cycloidal Crank Two-Gear Five-Bar Function Generator

According to the research (Reference 2) of A.V. Mohan Rao and G. N. Sandor, the synthesis of cycloidal crank two-gear five-bar linkage with the extension of Freudenstein's equation, the structural error of four points approximation is greater than that of the five points approximation of Freudenstein's four-bar linkage. It seems as if it is not worthy to use the geared linkage. On the contrary, it is worthwhile to do due to the fact that all the initial parameters as ϕ_1, γ_1, even ψ_1 are at the designer's own will, whereas in the synthesis of Freudenstein's four-bar linkage these parameters remain to be solved.

In general, the synthesis on cycloidal crank two-gear five-bar function generator may have got solutions as many as four mechanisms. However, questions should be considered that : Is it available? Does the ratio between the longest link to the shortest one suitable? Is it possible to reach every position prescribed? Besides, the limit and the dead center positions (Appendix G) should be considered for good operation conditions. After all, the extension of Freudenstein's equation is not a fine one for its long programming and complex procedure. Whereas the vector equation of closure is the simpler, especially for the number of precision positions is four or less as some linear equations are being served. Table 1.8 is the comparison between the two.

Table 1.8   Cycloidal Crank Two-Gear Five-Bar Function Generator

1.   Vector equation of closure

No. of precision positions	No. of algebrac equatios	Unknowns and their numbers	Arbitrarily chosen reals and their numbers	No. of unknowns to be solved for
$\Delta\phi, \Delta\psi^{*}$ prescribed; linear solution				
1	2 9	$\bar{L}_{21}, \bar{L}_{31}, \bar{L}_{41}, \bar{L}_{51}, n$	3 vectors of the left, n 7	2

2	4	Above $+\gamma_{12}$ 10	2 vectors of the left, n_1, γ_{12} 6	4
3	6	Above $+\gamma_{13}$ 11	1 vector of the left, n, γ_{12}, γ_{13} 5	6
4	8	Above $+ \gamma_{14}$ 12	$\boldsymbol{K_9}$, γ_{13}, γ_{14} 4	8
		$\Delta\phi, \Delta\psi$ prescribed; nonlinear solution		
5	10	Above $+ \gamma_{15}$ 13	n, γ_{12}, γ_{13} 3	10
6	12	Above $+ \gamma_{14}$ 14	n, γ_{12} 2	12
7	14	Above $+ \gamma_{17}$ 15	n 1	14
8	16	Above $+ \gamma_{18}$ 16	0	16
		Nothing prescribed, nonlinear solution		
9	18	Above $+ \gamma_{19}, \Delta\phi, \Delta\psi$ 19	n 1	18
10	20	Above $+ \gamma_{1\ 10}$ 20	0	20

* As $\Delta\phi$ and $\Delta\psi$ are the decision factor of scale factor R_ϕ and R_ψ when Δx and Δy are given, so the factors used are $\Delta\phi$ and $\Delta\psi$ instead of R_ϕ and R_ψ afterwards.

2. Extension of Freudenstein's equation

No. of precision positions	No. of algebraic equations	Unknowns and their numbers	Arbitrarily chosen reals and their numbers	No. of unknowns to be

				solved for
$\Delta\phi,\Delta\psi$, ϕ_1,ψ_1,γ_1,n prescribed; nonlinear solution				
1	1	L_2,L_3,L_4,L_5 4	3 of the left 3	1
2	2	Above 4	2 of the left 2	2
3	3	Above 4	1 of the left 1	3
4	4	Above 4	0	4
$\Delta\phi,\Delta\psi,n$ prescribed; nonlinear solution				
5	5	Above $+\phi_1,\psi_1,\gamma_1$ 7	ϕ_1,γ_1 2	5
6	6	Above 7	ϕ_1 1	6
7	7	Above 7	0	7
Nothing prescribed, nonlinear solution				
8	8	Above$+ \Delta\phi,\Delta\psi,n$ 10	$\Delta\phi,\Delta\psi$ 2	8
9	9	Above 10	n 1	9
10	10	Above 10	0	10

Notes (a) The first position is the initial position.

(b) The length of link 1 is equal to 1.

1.2   Four-Gear Five-Bar Linkage

1.2.1   Introduction

The four-gear five-bar linkage has more members than the ordinary ones, and its function is better too.

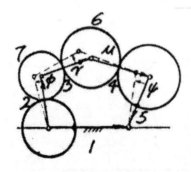

Figure 1.20   Four-gear five-bar linkage in its first (solid) and jth (dashed) precision positions

Now take the mechanism shown in Fig.1.20 as an example. First, calculate its degrees of freedom.

$$F = 3 \times (7-1) - 2 \times 7 - 1 \times 3 = 1$$

Next, find the angular displacements relationships among links. The following tables are to be tabulated for this purpose.

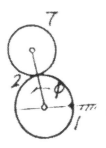

Figure 1.21   Input side of the four-gear five-bar linkage

Table 1.9

	Arm2	Gear 1	Gear 7
Motion with arm relation to frame	ϕ^{*}	ϕ	ϕ
Motion relative to arm	0	$-\phi$	$\dfrac{N_1}{N_7}\phi$
Total motion relative to frame	ϕ	0	$(1+\dfrac{N_1}{N_7})\phi$

* Sometimes substitute ϕ for ϕ_{1_j} due to simplification, and so are other angles.

The input side of this mechanism is an epicyclic gear train having link 2 (as an arm), gear 1 (as a frame), and gear 7, in Fig 1.21. If the input link 2 rotates an angle ϕ, gear 7 will rotate an angle $(1+\dfrac{N_1}{N_7})\phi$, in Table 1.9, where N_1 and N_7 are the numbers of teeth of gears 1 and 7 respectively.

The floating link 3 (as an arm) with gears 7, 6 compose another epicyclic gear train, Fig.1.22. Their rotations are shown in Table 1.10. Where the arm rotation angle γ is arbitrarily chosen.

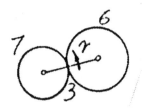

Figure 1.22   Floating link and idler gears

Table 1.10

	Arm3	Gear 7	Gear 6
Motion with arm relative to frame	γ	γ	γ
Motion relative to arm	0	$(1+\dfrac{N_1}{N_7})\phi-\gamma$	$[\gamma-(1+\dfrac{N_1}{N_7})\phi]\dfrac{N_7}{N_6}$
Total motion relative to frame	γ	$(1+\dfrac{N_1}{N_7})\phi$	γ + $[\gamma-(1+\dfrac{N_1}{N_7})\phi]\dfrac{N_7}{N_6}$

The final epicyclic gear train——link 4 (as an arm) with gears 6, 5——is the output side of the mechanism (Fig.1.23), whose angular displacement (via Table 1.11)

$$\psi = \mu + (\mu-\theta_6)\frac{N_6}{N_5}$$

$$= \mu + \{\mu-\gamma-[\gamma-(1+\frac{N_1}{N_7})\phi]\frac{N_7}{N_6}\}\frac{N_6}{N_5}$$

$$= \mu(1+\frac{N_6}{N_5})-\gamma(\frac{N_6+N_7}{N_5})+\phi(\frac{N_1+N_7}{N_5})$$

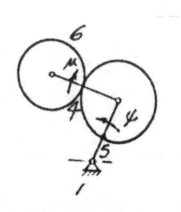

Figure 1.23   Output side of the four-gear five-bar linkage

Table 1.11

	Arm 4	Gear 6	Gear 5
Motion with arm relative to frame	μ	μ	μ
Motion relative to arm	0	$\theta_6^* - \mu$	$(\mu - \theta_6)\dfrac{N_6}{N_5}$
Total motion relative to frame	μ	θ_6	$\mu + (\mu - \theta_6)\dfrac{N_6}{N_5}$

$*\theta_6$ replaces $\gamma + [\gamma - (1 + \dfrac{N_1}{N_7})\phi]\dfrac{N_7}{N_6}$ for tabulating convenience.

Now the value of μ can be calculated by

$$\mu = \frac{1}{\left(1 + \dfrac{N_6}{N_5}\right)}\left[\psi + \gamma\left(\frac{N_6 + N_7}{N_5}\right) - \phi\left(\frac{N_1 + N_7}{N_5}\right)\right] \qquad (1.43)$$

where $\psi = f(\phi)$ is prescribed and γ is arbitrarily chosen. It is obvious that the number of teeth of each gear is the decisive factor in the synthesis on the mechanism.

1.2.2 Synthesis on Four-Gear Five-Bar Linkage

Since the four-gear five-bar linkage has one degree of freedom and one vector equation of closure, which can be written as follows:

$$e^{i\phi_{1j}}\overline{L}_{21} + e^{i\gamma_{1j}}\overline{L}_{31} + e^{i\mu_{1j}}\overline{L}_{41} + e^{i\psi_{1j}}\overline{L}_{51} = \overline{L}_{11} = 1$$

$$(1.44)$$

where $\overline{L}_{11}$ is assigned the value 1.

Assume the gear ratios, $\dfrac{N_1}{N_5}, \dfrac{N_6}{N_5}, \dfrac{N_7}{N_5}$, and the ranges $\Delta\psi$ and $\Delta\psi$ are prescribed; wholly or partly, the discussion about the number of precision positions of the synthesis on function generator can be discussed as follows (Table 1.12):

Table 1.12   Four-Gear Five-Bar Function Generator

No. of precision positions	No. of algebraic equations	Unknowns and their numbers	Arbitrarily chosen reals and their numbers	No. of unknowns to be solved for
$n_1(=\dfrac{N_1}{N_5}), n_2(=\dfrac{N_6}{N_5}), n_3(=\dfrac{N_7}{N_5}), \Delta\phi, \Delta\psi$ prescribed; linear solution			
1	2	$\bar{L}_{21}, \bar{L}_{31}, \bar{L}_{41}, \bar{L}_{51}$ 8	3 vectors of the left 6	2
2	4	Above + γ_{12} 9	2 vectors of the left + γ_{12} 5	4
3	6	Above + γ_{13} 10	1 vector of the left + γ_{12} , γ_{13} 4	6
4	8	Above + γ_{14} 11	γ_{12}, γ_{13}, γ_{14} 3	8
$n_1, n_2, n_3, \Delta\phi, \Delta\psi$ prescribed; nonlinear solution			
5	10	Above + γ_{15} 12	γ_{12} , γ_{13} 2	10
6	12	Above + γ_{16} 13	γ_{12} 1	12
7	14	Above + γ_{17} 14	0	14
n_1, n_2, n_3 prescribed; nonlinear solution			
8	16	Above + $\gamma_{18}, \Delta\phi, \Delta\psi$ 17	$\Delta\phi$ 1	16
9	18	Above + γ_{19}	0	18

		18		
Nothing prescribed; nonlinear solution				
10	20	Above + $n_1,n_2,n_3,$ $\gamma_{1\ 10}$ 22	n_1,n_2 2	20
11	22	Above + $\gamma_{1\ 11}$ 23	n_1 1	22
12	24	Above + $\gamma_{1\ 12}$ 24	0	24

The gear ratios being discussed belong to compound gear trains. If the simple gear train is used, restrictions will occur. For instance as in Fig.1.20, the restrictions are

$$\frac{N_6 + N_7}{N_1 + N_7} = \frac{L_{31}}{L_{21}} \qquad \text{and} \qquad \frac{N_6 + N_5}{N_1 + N_7} = \frac{L_{41}}{L_{21}}$$

$$(1.45)$$

Then the number of precision positions is limited to ten.

Supposing the four precision positions is preferred, and the three arbitrarily chosen reals are determined, the eight unknowns (i.e. $\overline{L}_{21}, \overline{L}_{31}, \overline{L}_{41}, \overline{L}_{51}$) can be just solved by the eight corresponding linear equations.

If the number of precision positions is over four, the compatibility equations must be added to yield a system for solving the parameters beside those arbitrarily chosen γ's. Of course it is very tedious. Now the synthesis equations of the four finitely separated precision position linkage shown are the same as eq.(1.13) which is used for inverted two-gear five-bar linkage. The only difference between the two mechanisms is the determination of parameter μ due to the number of gears.

There will be a solution spectrum appeared if the values of the various gear ratios, input and output scale factors, and $\gamma_{12}, \gamma_{13}, \gamma_{14}$ are known.

【Example 1.4】 Design a four precision position four-gear five-bar linkage generating function $y = x^2$, $0 \le x \le 1$.

Solution:

Let $\Delta\phi = 90^0, \Delta\psi = 90^0$

$$\gamma_{12} = 20^0, \gamma_{13} = 40^0, \gamma_{14} = 60^0$$

some parameters are same as example 1.1. i.e.

$$L_{1x} = 1.0 \qquad\qquad L_{1y} = 0$$

$$\phi_{12} = 24.35382^0 \qquad\qquad \psi_{12} = 8.443917^0$$
$$\phi_{13} = 58.795335^0 \qquad\qquad \psi_{13} = 42.885432^0$$
$$\phi_{14} = 83.149155^0 \qquad\qquad \psi_{14} = 83.149146^0$$

whereas by eq.(1.43), yields

$$\mu_{1j} = \frac{1}{1 + \dfrac{N_6}{N_5}}[\psi_{1j} + \gamma_{1j}(\frac{N_6 + N_7}{N_5}) - \phi_{1J}(\frac{N_1 + N_7}{N_5})]$$

and let

$$N_1 : N_5 : N_6 : N_7 = 1:1:1:\frac{1}{2},$$

then

$$\mu_{12} = \frac{1}{1+1}[8.443917^0 + 20^0(\frac{3}{2}) - 24.35382^0(\frac{3}{2})]$$
$$= 0.9565935^0$$

$$\mu_{13} = \frac{1}{2}[42.88543^0 + 40^0(\frac{3}{2}) - 58.795335^0(\frac{3}{2})] = 7.3425621^0$$

$$\mu_{14} = \frac{1}{2}[83.149146^0 + 60^0(\frac{3}{2}) - 83.149155^0(\frac{3}{2})] = 24.212707^0$$

By eq.(1.13) yields

$$L_{21x} = -0.5966870 \qquad\qquad L_{21y} = -1.562800$$
$$L_{31x} = 0.7061015 \qquad\qquad L_{31y} = 1.549840$$
$$L_{41x} = 1.2391740 \qquad\qquad L_{41y} = -0.97684$$
$$L_{51x} = -0.3485882 \qquad\qquad L_{51y} = 0.98481$$

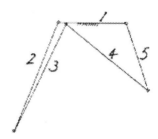

Figure 1.24 Four-gear five-bar function generator in its first precision position. $y = x^2, 0 \leq x \leq 1$, gears not shown.

It should be noted that the gear train has to be compounded for the meshing pairs

$$r_1 + r_7 = L_2 = 1.6728357$$
$$r_{7'} + r_6 = L_3 = 1.7031093$$
$$r_{6'} + r_5 = L_4 = 1.5779001$$

Where $r_1, r_7, r_{7'}, r_6, r_{6'}$ and r_5 are the radii of the pitch circles of gears 1, 7, 7', 6, 6' and 5. They are equal to 0.5576119, 1.1152238, 0.5677031, 1.1354062, 0.78895, and 0.78895 respectively.

2   Geared Four-Bar Linkages

2.1 Two-gear Four-Bar Linkages

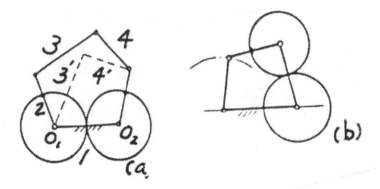

Figure 2.1 (a) The two-gear five-bar linkage is being transferred into two-gear
four-bar linkages;(b)the two-gear four-bar linkages inversion.

In fact, a two-gear five-bar linkage is a five member (i.e. link) mechanism because
link 2 and link 5 are rigidly attached to corresponding gears separately as an integral

member, Fig.2.1a. Let link 3 directly connect with the fixed joint O_1, the link portion

of the geared link 2 will be vanished, only the gear itself remains. Although the
number of links and the degrees of freedom of the linkage remain unchanged, the title
of it becomes two-gear four-bar linkage.

The two-gear four-bar linkage has three versions. The most useful one is shown in
Fig.2.1b, Being the basic mechanism of this geared linkage, the four-bar linkage can
be a crank rocker, a double crank, or a double rocker etc. We shall discuss one by one.

2.1.1   Crank Rocker mechanism as a basic mechanism

2.1.1.1 External Gear Pair

(1) Graphical Method

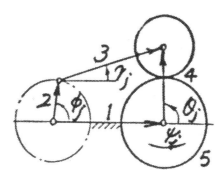

Figure 2.2   Two-gear four-bar linkage, crank rocker mechanism as a basic
mechanism

The epicyclic two-gear four-bar linkage with crank rocker as a basic mechanism is
shown in Fig.2.2.The numbers 1, 2, 3, 4, and 5 represent frame, crank, coupler, rocker,
and output gear; ϕ_j, γ_j, θ_j (θ takes the place of μ because link 4 is a rocker and
not a floating link now) and ψ_j as the jth position angles of crank, coupler, rocker
and gear respectively. And the following relationships should be mentioned.

$$\phi_j = \phi_0 + \phi_{oj}, \text{ let } \phi_0 = 0, \text{ then } \phi_j = \phi_{0j}$$

$$\gamma_j = \gamma_0 + \gamma_{0j}$$

$$\theta_j = \theta_0 + \theta_{0j}$$

$$\psi_j = \psi_0 + \psi_{0j}, \text{ let } \psi_0 = 0, \text{ then } \psi_j = \psi_{0j}$$

$$\left. \right\} \quad (2.1)$$

The geared link 3, link 4, and gear 5 composite a epicycle gear train with two degrees
of freedom, Since link 4 is the arm of this train, some relative relationships also
should be mentioned.

α_{r0} — initial position angle of link 3 relative to link 4

α_{rj} — jth position angle of link 3 relative to link 4

α_{r0j} — angular displacement of link 3 relative to link 4

$$\left. \begin{array}{l} \gamma_0 = \theta_0 + \alpha_{r0} \\ \gamma_j = \theta_j + \alpha_{rj} \end{array} \right\} \text{ see } \quad \text{Fig.2.3.} \quad \text{Then}$$

$$\gamma_{0j} = \theta_{0j} + \alpha_{r0j}$$

$$(2.2)$$

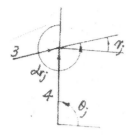

Figure 2.3 $\gamma_j = \theta_j + \alpha_{rj}$

$\alpha_{r0j} = \alpha_{rj} - \alpha_{r0}$

(2.3)

Contact with output angle,

$\psi_{0j} = \theta_{0j} + n\alpha_{r0j} = \theta_{0j} + n(\alpha_{rj} - \alpha_{r0})$ (2.4)

or $\psi_{0j} = \theta_{0j} + n(\gamma_{0j} - \theta_{0j})$ (2.5)

where $n = \pm\dfrac{N_3}{N_5}$, (negative sign is used here for external mesh)

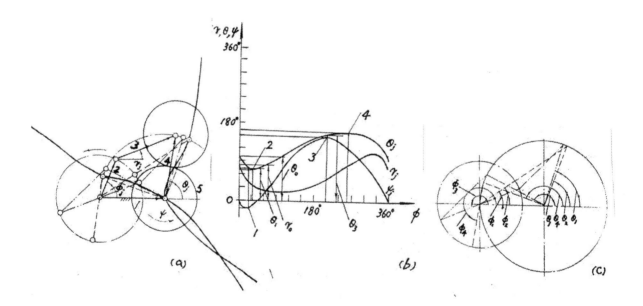

Figure 2.4 Two-gear four-bar linkage with crank rocker mechanism as a basic mechanism—non-reverted type:

$L_1 : L_2 : L_3 : L_4 = 1.5 : 1 : 1.5 : 1.5$; $r_3 : r_5 = 1 : 1$; $n = -1; \phi_0 = 0, \psi_0 = 0$.

(a) configuration and the centrode   13

(b) angular displacement curves γ, θ, ψ of links 3, 4, and gear5;

(c) some key angles.

Notes in (b)   in (c)

$$1 \quad \underline{\quad\quad} \quad \left(\phi_{\phi\min}_{22^0}, \begin{cases} \psi_{\min} \\ {}_{-24^0} \\ \theta_{\psi\min} \\ {}_{73.1^0} \end{cases} \right) \qquad\qquad \begin{array}{l} \phi_1 \underline{\quad} \phi_{\psi\min}_{22^0} \\[1em] \theta_1 \underline{\quad} \theta_{\psi\min}_{73.1^0} \end{array}$$

$$2 \quad \underline{\quad\quad} \quad (\phi_{\theta\min}_{34.5^0}, \theta_{\min}_{69.5^0}) \qquad\qquad \begin{array}{l} \phi_2 \underline{\quad} \phi_{\theta\min}_{34.5^0} \\[1em] \theta_2 \underline{\quad} \theta_{\min}_{69.5^0} \end{array}$$

$$3 \quad \underline{\quad\quad} \quad \left(\phi_{\psi\max}_{202^o}, \begin{cases} \psi_{\max} \\ {}_{149.5^0} \\ \theta_{\psi\max} \\ {}_{153.5^0} \end{cases} \right) \qquad\qquad \begin{array}{l} \phi_3 \underline{\quad} \phi_{\psi\,m\,a\,)}_{2\,0^0 2} \\[1em] \theta_3 \underline{\quad} \theta_{\psi\,m\,a\,:}_{1\,5.5^0} \end{array}$$

$$4 \quad \underline{\quad\quad} (\phi_{\theta\max}_{259^0}, \theta_{\max}_{161^0}) \qquad\qquad \begin{array}{l} \phi_4 \underline{\quad} \phi_{\theta\max}_{259^0} \\[1em] \theta_4 \underline{\quad} \theta_{\max}_{161^0} \end{array}$$

$$\gamma_0 = 80^0$$
$$\theta_0 = 101^0$$

Fig.2.4 show the configuration and curves of angular displacement of members vs the input crank rotation angle (i.e. time for the crank rotates with constant speed) of the mechanism of non-reverted type. Besides, somewhere key angles such as the dead center and limit positions are designated. The initial positions of input and output members, ϕ_0 and ψ_0, are all equal to zero, $\psi_{0j} = 2\theta_{0j} - \gamma_{0j}$ may become

$\psi_j = 2\vartheta_{0j} - \gamma_{0j}$. Since θ_{0j} and γ_{0j} may be drawn graphically by every increment

of crank angle if needed, then the output data ψ_j will be obtained by this equation. It

should be noted that at curve ψ_j's valley point (i.e. the limit position of gear 5)

where $\phi_{\psi\min} = 22^0$, link 4 still goes on until it reaches at the point where

$\phi_{\theta\min} = 34.5^0$. And at the crest point $\phi_{\psi\max} = 202^0$, the another limit position of gear

5; link 4 goes on also until it reaches at $\phi_{\theta\max} = 259^0$. Obviously, this phenomenon

shows the existing difference of limit positions between the two members 4, 5. The limit position of gear 5 occurs when the instantaneous certer 13 of links 1, 3 coincides with that of 35 of links 3, 5 (simultaneously 12 and 25). The path of 35 does possess the left upper portion of the pitch circle circumference of gear 5. The trace of 13 intersects with the trace of 35 at two points when $\phi_{\psi \min}$ and $\phi_{\psi \max}$ occur . The limit position of link 4 occurs when (i) links 2, 3 line up as a straight line when $\phi_{\theta \min}$ exists, (ii) links 2, 3 coincides each other when $\phi_{\theta \max}$ exists.

Observing the curves in Fig.2.4b, We know:
(1) All the moving members except link 2 are oscillating ones.
(2) The oscillating amplitude of gear 5 is greater than that of link 4, i.e.

$$\psi_{\max} - \psi_{\min} = 149.5^0 - (-24^0) = 173.5^0$$
$$\theta_{\max} - \theta_{\min} = 161^0 - 69.5^0 = 91.5^0$$

(3) $\phi_{\psi \max} - \phi_{\psi \min} = 202^0 - 22^0 = 180^0$

It indicates that the mechanism is no longer a quick return mechanism since gear 5 replaces link 4 (which was a quick return member of a conventional crank rocker mechanism) as an output member now. And the difference between $\psi_{\max}$ and

$\psi_{\min}$ is just 180^0 of the input displacement.

There leaves a title " non-reverted type " to be explained. Assume the geometrical center line of the output shaft coincides with that of the input shaft, as if the power flow reverted to the input end. This is why the reverted type is named. On the contrary, if the power flow did not " revert" to the input end, then the " non-reverted type" would be entitled.
Now a reverted type will be discussed and is shown in Fig.2.5 which the center line of output gear 5 revolves multiple revolutions as the input crank revolves one revolution. The rule is

$$\psi_{0j} = \phi_{0j} + n(\gamma_{0j} - \phi_{0j}) \tag{2.6}$$

then $\psi_j = 2\phi_j + \gamma_0 - \gamma_j$; for $\psi_0 = 0$, $\phi_0 = 0$, and $n = -1$

$$\psi_j = \frac{3}{2}\phi_j + \frac{1}{2}(\gamma_0 - \gamma_j); \text{if } n = -\frac{1}{2} \text{ (i.e. } N_3 \Big/ N_5 = 1 \Big/ 2 \text{)}$$

From Fig.2.5b and refer to the above equations, we know that the transmission ratio between the output and input shafts is obeyed only when every moving cycle is completed:such as being 2 when $n = -1$; $3 \Big/ 2$ when $n = -\frac{1}{2}$. The output shaft oscillates no longer, which differs from the non-reverted type distinctly

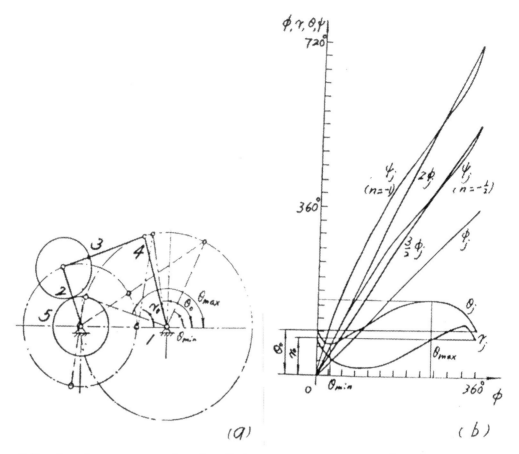

(a)   (b)

Figure 2.5   Another two-gear four-bar linkage with crank rocker mechanism as a basic mechanism—reverted type. $L_1 : L_2 : L_3 : L_4 = 1.5 : 1 : 1.5 : 1.5$; $r_3 : r_5 = 1 : 1$;

n=-1; $\phi_0 = 0$, $\psi_0 = 0$.

(a)   configuration; (b) angular displacement curves of links 2, 3, 4 and gear 5; in addition, a ψ_j curve with $n = -\dfrac{1}{2}$ is presented.

(2) Analytical Method

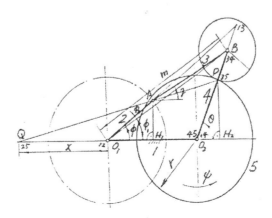

Figure 2.6   Two-gear four-bar linkage with crank rocker mechanism as a basic

mechanism — non-reverted type
Draw all the required instantaneous centers of the mechanism, Fig.2.6 . Let the input
crank 2's angular velocity is constant $\dot{\phi}$, the output gear 5's angular velocity is $\dot{\psi}$. If
the distance from the instantaneous center 25 (i.e. point Q) to the rotating center O_1
of link 2 is x , the length of link 1 is L_1 , then

$$\frac{\dot{\psi}}{\dot{\phi}} = \frac{x}{x+L_1} \tag{2.7}$$

Draw two vertical lines separately from instantaneous centered 23 (i.e. point A) and
35 (i.e. point P) to intersect the center line of the frame 1 at points H_1 and H_2. Let
r be the pitch circle radius of gear 5, then

$$\Delta PQH_2 \sim \Delta AQH_1$$

yields

$$\frac{r\sin\theta^*}{L_2\sin\phi} = \frac{x+L_1+r\cos\theta}{x+L_2\cos\phi}$$

$$x = \frac{L_2[L_1\sin\phi - r\sin(\theta-\phi)]}{r\sin\theta - L_2\sin\phi} \tag{2.8}$$

From eqs.(2.7), (2.8), we have

$$\frac{\dot{\psi}}{\dot{\phi}} = \frac{L_2[L_1\sin\phi - r\sin(\theta-\phi)]}{\gamma[L_1\sin\theta - L_2\sin(\theta-\phi)]} \tag{2.9}$$

Integrating eq. (2.9) gives

$$\psi = \int \frac{L_2[L_1\sin\phi - r\sin(\theta-\phi)]}{r[L_1\sin\phi - L_2\sin(\theta-\phi)]}d\phi \tag{2.10}$$

This equation is the angular displacement of output angle ψ versus input angle ϕ.

Whereas the relationship between ϕ and θ, see Appendix H.

* θ is instead of θ_j just for simplicity if no confusion is made. And so is the ϕ etc

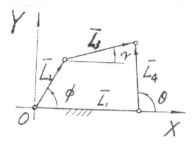

Figure 2.7   A typical planar four-bar linkage

The angular velocities among links of planar four-bar linkage have been fully introduced in traditional text books. As for comparison, we take a equation from the text book, that is

$$\dot{\theta} = \frac{L_2 \sin(\phi - \gamma)}{L_4 \sin(\theta - \gamma)} \dot{\phi} \qquad (2.11)$$

But considering these angles in Fig.2.7 are not easy to treat, so try to use the Freudenstein's displacement equation, Appendix E, we have

$$\theta = 2 \tan^{-1} \frac{D \pm \sqrt{D_2 + E^2 - F^2}}{E + F}$$

Where θ is ψ in Freudenstein's equation, and

$$D = \sin\phi, \quad E = -\frac{L_1}{L_2} + \cos\phi,$$

$$F = -\frac{L_1}{L_4} + \cos\phi + \frac{L_1^2 + L_2^2 - L_3^2 + L_4^2}{2 L_2 L_4}$$

Since all the lengths of the links are known, and the function $\theta = f(\phi)$ is a single value function, so this is preferable.

If the extremes of ψ is required, we may use eq.(2.9). Let $\dot{\psi} = 0$,

i.e. $[L_1 \sin\phi - \gamma \sin(\theta - \phi)] = 0$.

Also, it is advisable to use the coincidence of instantaneous centers 13 and 35. Draw a link along link 2 passing through the pitch point 35 between the two gears. See Fig 2.8.

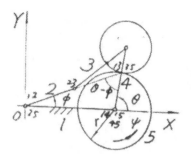

Figure 2.8 The extremes of output angle ψ of the non-reverted two-gear four-bar linkage with crank rocker mechanism as a basic mechanism

2.1.1.2 Internal Gear Pair

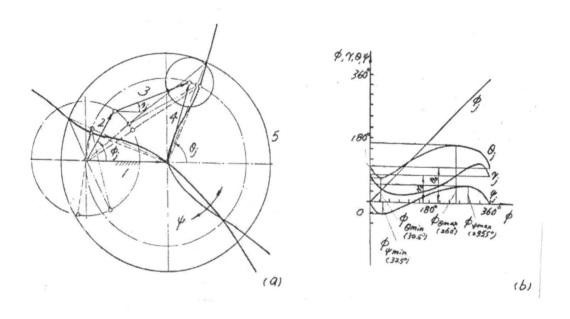

(a)

(b)

Figure 2.9 Non-reverted two-gear four-bar mechanism with crank rocker mechanism as a basic mechanism, internal gear pair, $N_3/N_5 = 1/4$. (a) Configuration with the centrode 13 ; (b) angular displacement curves.

Take an internal gear pair instead of the external pair of gears 3, 5 of the mechanism, and let $r_3 : r_5 = 1:4$. The oscillating angle $(\psi_{max} - \psi_{min})$ will be no longer greater but less than that ($\theta_{max} - \theta_{min}$) of corresponding link, however. And in this example

$$\psi_j = \theta_{0j} + \frac{1}{4}(\gamma_{0j} - \theta_{0j}) = \frac{3}{4}(\theta_j - \theta_0) + \frac{1}{4}(\gamma_j - \gamma_0)$$

2.1.2   Double Crank Mechanism as a Basic Mechanism

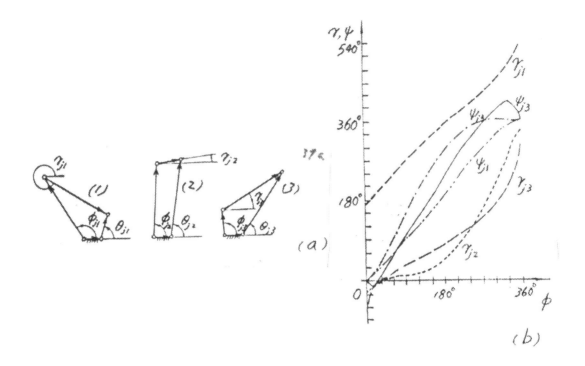

Figure 2.10   Reverted two-gear four-bar linkage with unequal length drag link mechanism as a basic mechanism . (a) Configurations 1, 2, 3; n= -1, gears not shown;

(b) angular displacement curves of cycloidal crank and output gear, γ and ψ .

The equal length double crank four-bar linkage (i.e. the parallelogram four-bar linkage) is the simplest of the drag link mechanism; whereas the unequal length double crank four-bar linkage, however, has general significance. From Fig.2.10b, we know that there are three different cases possibly caused by different relative dimensions as follows:

(1) non-stop, irreversible, going on with varied velocity, ψ_{j1};

(2) irreversible, only stop once (velocity being zero), going on instantly, ψ_{j2};

(3) reversing partly, stop twice (velocity being zero at the beginning and the end of
   reversing interval), ψ_{j3}.

It may be nice to take an inverted equilateral trapezoid double crank mechanism as a basic mechanism for convenience. Having observed the two examples in Fig.2.11a, we define a typical form, namely $L_1 : L_2 : L_3 : L_4 = 1 : 1.5 : 1.5 : 1.5$, shown in

Fig.2.12a.   Varying the value of n, we can obtain many different ψ curves. Let

$\gamma_0 = O^0$ be the initial position of the cycloidal crank, just for comparison convenience. While the inverted equilateral trapezoid double crank mechanism is used as a basic mechanism, not only the output gear makes a complete revolution and with the same direction as the input crank does, but also the link 4 makes a complete revolution.

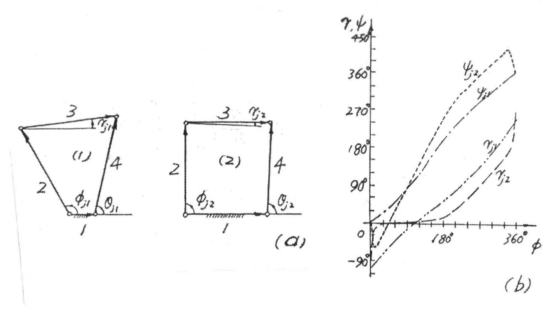

Figure 2.11 Reverted two-gear four-bar linkage with inverted equilateral trapezoid double crank mechanism as a basic mechanism. (a) Configurations 1, 2; $n = -1$, gears not shown, (b) angular displacement curves of cycloidal crank and output link.

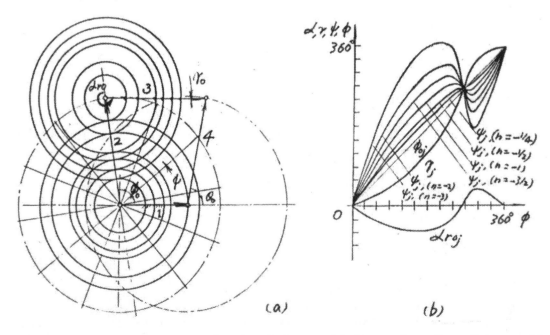

Figure 2.12 Reverted two-gear four-bar linkage with inverted equilateral trapezoid double crank mechanism as a basic mechanism.
(a) Configurations;

for $\qquad n = -\dfrac{N_3}{N_5} = -\dfrac{1}{4}, -\dfrac{1}{2}, -1, -\dfrac{3}{2}, -2, -3, \psi_0 = 0, \gamma_0 = 0$,

$\gamma_j = \gamma_{0j}, \alpha_{\gamma 0 j} = \gamma_j - \phi_{0j}$;

(b) corresponding output angular displacement curves.

Fig.2.13 shows a non-reverted mechanism. We know that the output angular displacement curve ψ_j has a crest and a valley at about 235^0 and 318^0 of input crank angle respectively. These are the two extremes

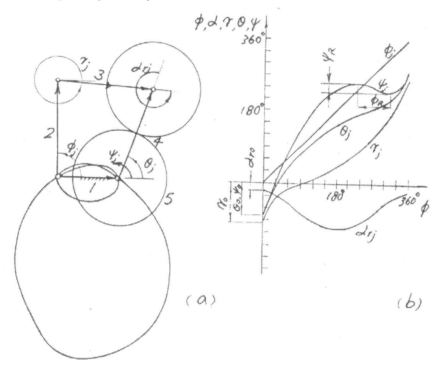

(a)　　　　　　　　　　　　(b)

Figure 2.13　　Non-reverted two-gear four-bar linkage with inverted equilateral trapezoid double crank mechanism as a basic mechanism. (a) Configuration and instantaneous center trace of links 1, 3; $\phi_0 = 0, \theta_0 = \psi_0, \alpha_{rj} = \gamma_j - \theta_j, n = -1$; (b) curves of angular displacement.

points where the instantaneous dwells occur for velocities being zeros. While the direction of the angular displacement is reversed between these two points, the so called " reversing phenomenon " presents. In this example, the reversing angle $\psi_R (\approx 24^0)$, the reversing time equivalent to the input crank rotation angle $\phi_R (\approx 318^0 - 235^0 = 83^0)$ can be measured.

Now we review back to Fig 2.12 b, there are " reversing phenomena " too. We take the curve of n = -1 as a comparison. The measured data are: $\psi_R \approx 26^0, \phi_R \approx 278^0 - 240^0 = 38^0$. Both the location and the magnitude are quite different. No doubt, this is a good omen. Studying the phenomena fully, we can utilize it for dwelling purpose because the ψ_R can be as small as we need within the tolerance.

Although the simple dwell mechanism such as ratchets, Geneva wheels are widely used in engineering, their functions can not satisfy the requirements of developing industry. Therefore, this new kind of dwell mechanism should be studied on schedule. Certainly, the above discussion can be applied with analytical method, but it would be neglected here

.2.1.2.1   The General Reverted Type Discussion

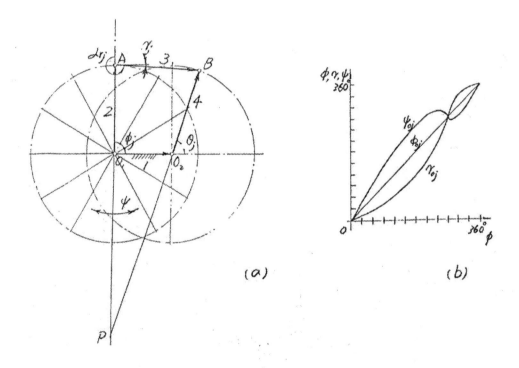

Figure 2.14   A typical reverted two-gear four-bar linkage with inverted equilateral trapezoid double crank mechanism.(a) Configuration $n = -1$, gears not shown;(b) angular displacement curves

We have already discussed the inverted equilateral trapezoid as a basic mechanism for the double crank series on both   reverted and non-reverted types. It is necessary, however, to study about its general characteristics further. And the reverted type is preferred by chance. As we all know that the output angular displacement ψ_{oj} depends

upon   input   link's   angular   displacement ϕ_{oj} ,relative   length ratio $\left(i.e. L_2/L_1, L_3/L_1, L_4/L_1\right)$, and $n\left(= \pm\dfrac{N_3}{N_5}\right)$. We see in Fig.2.14a that ψ_{oj} is the superposition of ϕ_{oj} and $n\alpha_{roj}$, that is

$$\psi_{oj} = \phi_{oj} + n\alpha_{roj}$$

where α_{roj} is the angular displacement from initial position to jth position of gear 3 which is rigidly attached to link 3 as an integer relative to link 2, so it can be equal to $\gamma_{oj} - \phi_{oj}$, then

$$\psi_{oj} = \phi_{oj} + n\left(\gamma_{oj} - \phi_{oj}\right), \text{just the same as eq.(2.6)}.$$

For simplicity, write ϕ, ψ, α_r and γ instead of $\phi_{oj}, \psi_{oj}, \alpha_{roj}$, and γ_{oj}. Hence eq.(2.6) will be

$$\psi = \phi + n\left(\gamma - \phi\right)$$

(2.12)

$$\text{or} \qquad\qquad = \phi + n\alpha_r$$

By differentiating eq.(2.12) with respect to time, and rearranging it, we get

$$\frac{\dot{\psi} - \dot{\phi}}{\dot{\gamma} - \dot{\phi}} = n$$

Where $n = -\dfrac{N_3}{N_5}$, then

$$\dot{\psi} = \left(1 - n\right)\dot{\phi} + n\dot{\gamma} \text{ ; and for } \dot{\gamma} = \dot{\phi} + \dot{\alpha}_r$$

Let

$$\lambda = \frac{\dot{\psi}}{\dot{\phi}} = 1 - n + n\frac{\dot{\gamma}}{\dot{\phi}} = 1 - \frac{n\left(\dot{\phi} - \dot{\gamma}\right)}{\dot{\phi}}$$

$$= 1 + n\frac{\dot{\alpha}_r}{\dot{\phi}}$$

(2.13)

where $\dot{\phi}$ is a constant.

The ratio $\dfrac{\dot{\alpha}_r}{\dot{\phi}}$ is hard to calculate unless link 2 is assumed to be a "frame" in imagination. Then the ratio $\dot{\alpha}_r/\dot{\phi}_r$ will be easy to obtain. Since the $\dot{\alpha}_r$ and $\dot{\phi}_r$ are just the output and input angular velocities of the "four-bar linkage" respectively. Where $\dot{\phi}_r$ is the angular velocity of link 1 relative to link 2, just the reversal of $\dot{\phi}$, i.e. $\dot{\phi}_r = -\dot{\phi}$.

From Appendix H, we have

$$\frac{\dot{\phi}}{\dot{\theta}} = \frac{L_4}{L_2}\left[\frac{L_1\sin\theta - L_2\sin(\theta-\phi)}{L_1\sin\phi - L_4\sin(\theta-\phi)}\right]$$

Replacing $\dot{\phi},\dot{\theta},L_1,L_2,L_4,\phi,$ and θ by $\dot{\alpha}_r,\dot{\phi}_r,L_2,L_3,L_1,\alpha_r - 180^0$, and $180^0 - \phi$ respectively,

yield
$$\frac{\dot{\alpha}_r}{\dot{\phi}_r} = \frac{L_1}{L_3}\left[\frac{L_2\sin\phi + L_3\sin(\alpha_r+\phi)}{-L_2\sin\alpha_r + L_1\sin(\alpha_r+\phi)}\right]$$

(2.14)

In addition,

$$\frac{\dot{\alpha}_r}{\dot{\phi}_r} = \frac{\overline{O_1P}}{\overline{AP}}$$

(2.15)

where the directions of $\overline{O_1P}$ and $\overline{AP}$ determine the sense of the ratio $\dot{\alpha}_r/\dot{\phi}_r$; positive if the two vectors are in the same direction, otherwise negative. Substituting eq.(2.14) into eq.(2.13), we have

$$\lambda = 1 - n\left(\frac{L_1}{L_2}\right)\left[\frac{L_3\sin(\alpha_r+\phi)+L_2\sin\phi}{L_1\sin(\alpha_r+\phi)-L_2\sin\alpha_r}\right]$$

(2.16)

Substituting eq.(2.15) into eq.(2.13), we have

$$\lambda = 1 - n\frac{\overline{O_1P}}{\overline{AP}}$$

(2.17)

If the angular velocity of the output pinion 5, $\dot{\psi} = 0$, then $\lambda = 0/\dot{\phi} = 0$. We define it as first order dwell. From eq.(2.17), yields

$$\frac{\overline{O_1P}}{\overline{AP}} = \frac{1}{n}$$

(2.18)

This equation expresses that the instantaneous enters 13 and 35 coincide, i.e. the intersecting point of "frame" and "coupler" coincides with the common tangent (i.e. the pitch point) of the pitch circles of the two gears (Fig.2.15)

If the angular acceleration of pinion 5, $\ddot{\psi} = 0$ occurs at the same instant which will

be defined as second order dwell, i.e. $\ddot{\psi} = \dot{\phi}\dfrac{d\lambda}{dt} = n\dfrac{d^2\alpha_r}{dt^2} = 0$, then

$$\frac{d^2\alpha_r}{dt^2} = 0$$

(2.19)

Equation(2.19) means that the "output link" 3's angular acceleration $\ddot{\alpha}_r$ is zero while the "input link"1 is rotating with a constant angular speed.

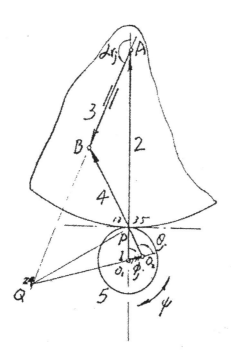

Figure 2.15 Reverted two-gear four-bar linkage with double crank mechanism as a basic mechanism when both the angular velocity $\dot{\psi}$ and acceleration $\ddot{\psi}$ of output are equal to zero

The Freudenstein's Theorem states that: In a four-bar mechanism, in the phase corresponding to the extreme values of the velocity ratio, the collineation axis* is perpendicular to the coupler. That is, the second order dwell presents; while $\dot{\phi}_r$ =constant, $\ddot{\alpha}_r$ =0

Now look at the mechanism in Fig.2.15. The intersecting point of the "frame" and the "coupler" coincides with the pitch point of the pitch circles of the two gears, first order dwell; the collineation axis PQ is perpendicular to "coupler", second order dwell. This is an example of both the angular velocity and acceleration of output member are equal to zeros, however, it is not an inverted equilateral trapezoid basic mechanism unless links 2,4 are equal in length and then the instantaneous center 24 will be in infinity.

*In a four-bar linkage, the line connecting the relative instantaneous centers 13 and 24 of the physically separate links is called the collineation axis of the mechanism.

2.1.2.2   One Special Reverted Type

—External Planetary System(Epicyclic Gear Train) with Fixed Eccentric Control Mechanism

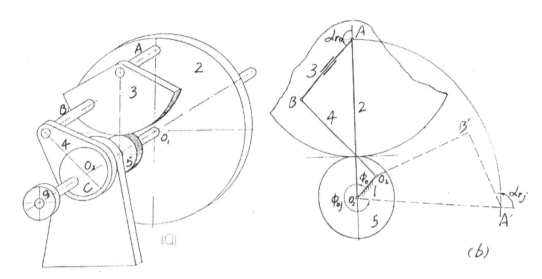

Figure 2.16   Epicyclic gear train with fixed control mechanism:(a) oblique drawing;(b)configurations of the mechanism in the initial(solid) and jth(dashed)positions

Fig.2.16 shows a special reverted type of two-gear four-bar linkage with double crank mechanism as a basic mechanism. Flywheel 2 is attached to the input shaft's geometric center O_1. Gear segment 3(i.e.member *AB)* oscillates about the pivot at A of flywheel 2 and meshes with the output pinion 5. Link 4 rotates about axis C whose center is O_2, which is rigidly attached to frame and eccentric to O_1.Being the input link, flywheel 2 rotates uniformly and transmits force to gear segment 3 and then to link 4. Link 4 rotates axis without output or energy consumed if friction is neglected. The pinion 5 revolved by the gear segment 3 is just the output menber for transmitting energy to its receivers(not shown in Fig.2.16).

The number of degrees of freedom of this mechanism

F=3(5-1)-2×5-1×1=1

Assuming that in the initial position the intersecting point of "frame" AO_1 and "coupler" BO_2 coincides with the pitch point of the two pitch circles of the two gears, then the angular velocity of the output pinion 5,$\dot{\psi} = 0$. And $\ddot{\psi} = 0$ also if AB is perpendicular to "BO_2 and parallel to O_1O_2(the special case of the colineation axis perpendicular to "coupler"), meanwhile link 2 rotates clockwise with a uiform angular velocity ω to jth position.

The motion of output pinion:

$$\left.\begin{array}{l}\psi_{oj} = \phi_{oj} + n\alpha_{roj} \\ (i.e.\psi = \phi + n\alpha_r) \\ \dot{\psi} = \omega + n\dot{\alpha}_r \\ \ddot{\psi} = n\ddot{\alpha}_r\end{array}\right]$$

(2.20)

The angular displacement of planetary gear segment relative to arm AO_1:

$$\alpha_r = \alpha_{roj} = \alpha_{rj} - \alpha_{ro} = \pi - \alpha_{ro} - (\pi - \alpha_{rj})$$

$$= \pi - \alpha_{ro} - \left(\tan^{-1} \frac{L_1 \sin(2\pi - \phi_j)}{L_2 - L_1 \cos(2\pi - \phi_j)} + \cos^{-1} \frac{L^2 + L_3^2 - L_4^2}{2L_3 L} \right)$$

(2.21)

where $L_1 = \sqrt{L_1^2 + L_2^2 - 2L_1 L_2 \cos(2\pi - \phi_j)}$

$$2\pi - \phi_j = \pi - (\phi_{oj} - \alpha_{ro})$$

Then

$$\dot{\alpha}_r = \frac{d\alpha_{roj}}{dt}$$

$$= \omega L_1 \left[\frac{L_2 \cos(2\pi - \phi_j) - L_1}{L^2} - \frac{L_2 \sin(2\pi - \phi_j)(L^2 - L_3^2 + L_4^2)}{L^2 \sqrt{4L_3^2 L^2 - (L^2 + L_3^2 - L_4^2)^2}} \right]$$

(2.22)

$$\ddot{\alpha}_r = \frac{d^2 \alpha_{roj}}{dt^2}$$

$$=$$

$$-\omega^2 L_1 L_2 \left\{ \frac{1}{\sqrt{4L_3^2 L^2 - (L^2 + L_3^2 - L_4^2)^2}} \left[\frac{2L_1 L_2 \sin^2(2\pi - \phi_j)(L_3^2 - L^2)}{L^4} + \frac{L^2 - L_3^2 + L_4^2}{L^2} \left(\cos(2\pi - \phi_j) - \frac{2L_1 L_2 \sin^2(2\pi - \phi_j)(L_3^2 - L^2 + L_4^2)}{4L_3^2 L^2 - (L^2 + L_3^2 - L_4^2)^2} \right) \right] + \frac{\sin(2\pi - \phi_j)}{L^2} \left(\frac{2L_1 L_2 \cos(2\pi - \phi_j) - 2L_1^2}{L^2} + 1 \right) \right\}$$

(2.23)

【Example 2.1】 Prove that the epicyclic gear train and the fixed eccentric control mechanism at the beginning($\phi_0 = 45^0$) and the end ($\phi_f = 405^0$) of the motion cycle;

the output pinion's angular velocities $\dot{\psi}_0, \dot{\psi}_f$;and angular accelerations $\ddot{\psi}_0, \ddot{\psi}_f$ are

all equal to zero;

$L_1 = r_5 \cos 45^0, L_3 = r_3 \cos 45^0, L_4 = L_1 + L_3 = 0.707L_2, r_5 = 1, r_3 = 2, \alpha_{r0} = 135^0$,and the

angular velocity of link 2 is constant

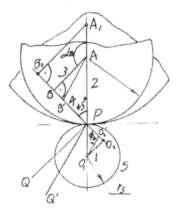

Figure2.17 Example of the e antpicyclic gear train and fixed eccentric control

mechanism. $\dot{\psi}_0 = \ddot{\psi} = 0, \dot{\psi}_f = \ddot{\psi}_f = o$;when $\phi_0 = 45^0, \phi_f = 405^0$ .The basic four-bar

linkage is $O_2 O_1 AB$.

Solution:

(1) Graphical method

Since at the beginning($\phi_0 = 45^0$) and the end($\phi_f = 405^0$) the " coupler" and the

"frame" intersects at the common tangent of the two gear's pitch circles which
meets the first order dwell condition, so both the angular velocities of output

pinion $\dot{\psi}_0, \dot{\psi}_f$ are equal to zero.

Further-more according to Freudenstein's theorem both the $\ddot{\psi}_0, \ddot{\psi}_2$ are equal to

zero for the collineation axis is perpendicular to the "coupler". If the basic

linkage is replaced by $AB'O'_2 O_1 (AB'//O'_2 O_1)$, The collineation axis is no longer

perpendicular to the "coupler", then $\ddot{\psi}$ will not equal to zero (see Fig.2.17).

(2) Analytical method

For $2\pi - \phi_0 = 315^0, 2\pi - \phi_f = -45^0$; these two are equivalent.

Substituting relevant values into eq.(2.22) yields

$$\dot{\alpha}_r = \omega\left(\frac{\sqrt{2}}{2}\right)\left[\frac{3\left(\frac{\sqrt{2}}{2}\right) - \frac{\sqrt{2}}{2}}{6.5} + \frac{3\left(\frac{\sqrt{2}}{2}\right)\left(\frac{9}{6.5}\right)}{\sqrt{(4)(2)(6.5) - 16}}\right] = \frac{1}{2}\omega$$

By eq. (2.20) yields

$$\dot{\psi} = \omega + n\dot{\alpha}_r = \omega + (-2)\left(\frac{\omega}{2}\right) = 0$$

And

$$\ddot{\alpha}_r = -\omega^2\left(\frac{\sqrt{2}}{2}\right)(3)\left\{\frac{1}{\sqrt{36}}\left[\left[\frac{2\left(\frac{\sqrt{2}}{2}\right)(3)\left(\frac{1}{2}\right)(2-4.5)}{6.5^2} + \left(\frac{6.5-2+4.5}{6.5}\right)\right]\left[\frac{\sqrt{2}}{2} - \frac{(2)\left(\frac{\sqrt{2}}{2}\right)\left(\frac{1}{2}\right)(2-6.5+4.5)}{36}\right]\right] + \frac{-\frac{\sqrt{2}}{2}}{6.5}\left[(2)\left(\frac{\sqrt{2}}{2}\right)\left(\frac{3\sqrt{2}}{2} - \frac{\sqrt{2}}{2}\right)\left(\frac{1}{6.5}\right) + 1\right]\right\}$$

$$= -\frac{3}{2}\sqrt{2}\omega^2\left\{\frac{51\sqrt{2} - 51\sqrt{2}}{(6)(6.5)(6.5)(2)}\right\}$$

=0

Therefore $\ddot{\psi} = n\ddot{\alpha}_r = 0$

Warning :It should be noticed that although this mechanism can meet the requirement of proof,it fails in operation for its disobeying of the Grashoff's Law. However, it can be operated if $r_3 : r_5 = 3 : 1$ is chosen. The configuration

$O_2 O_1 AB$ may become $O_2 O_1 A_1 B_1$. Unfortunately,the problem thus solved is not

satisfactory for the ratio of L_2 to L_1 is too big. If we add eq.(2.25)(Grashoff's

Law's limit) to eq.(2.24)which prescribed already as below,we may get a result though it is still not very good.

$$\left.\begin{array}{l} (L_1 + L_3 + L_4)\cos 45^0 = L_2 \\ L_4 = L_1 + L_3 \\ 2L_1 \cos 45^0 = 1 \end{array}\right\} \tag{2.24}$$

$$L_1 + L_2 \leq L_3 + L_4$$

(2.25)

2.1..2.3   Another of the Special Reverted Type

------   Sliding   rack   with   fixed   eccentric   control   mechanism

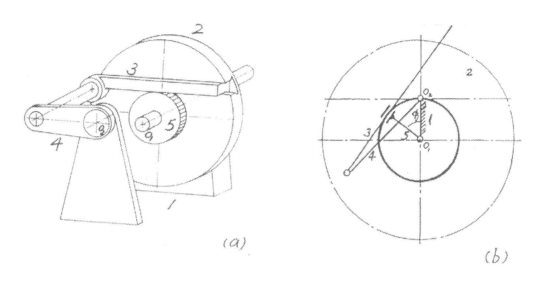

(a)

(b)

Figure   2.18   Sliding rack with   fixed eccentric control mechanism: (a) oblique drawing; (b) configuration of the mechanism in the jth position

The number of degrees of freedom of the sliding rack and fixed eccentric control mechanism.

$$F=3(5-1)-2\times5-1\times1=1$$

In Fig.2.18 rack 3 replaces gear segment as a coupler sliding in the groove of flywheel 2 and meshing with

output pinion 5. Link 4 rotates about O_2 which is eccentric $(=L_1)$ to O_1. If the driving member 2 rotates counterclockwise with a uniform angular velocity ω, the output pinion radius r_5 is equal to the fixed eccentric O_1O_2 and

L_5 (i.e. $r_5 = L_1 = L_5$),while the effective length of L_3 is not certain. Thus the output pinion may present first   and second order dwell both at the beginning and the end of a periodical motion; i.e. $\dot{\psi}_{\phi o} = 0, \ddot{\psi}_{\phi o} = 0, \dot{\psi}_{\phi f} = 0, \ddot{\psi}_{\phi f} = 0$.

The rolling of sliding rack relative to pinion;

$$s_r = L_2 + r_5 \sin \phi^* - \sqrt{L_2^{\,2} - r_5^{\,2}(1 - \cos \phi)^2}$$

(2.26)

*It is simpler than ϕ_j if no confusion from Y-axis as the base-line is made. The same is ψ and so on.

$$v_r = r_5 \omega \left[\cos \phi + \frac{\sin \phi (1 - \cos \phi)}{\sqrt{\dfrac{L_2^{\,2}}{r_5^{\,2}} - (1 - \cos \phi)^2}} \right]$$

(2.27)

$$a_r = r_5 \omega^2 \left\{ -\sin \phi + \frac{1}{\sqrt{\dfrac{L_2^2}{r_5^{\,2}} - (1 - \cos \phi)^2}} \left[\sin^2 \phi + \cos \phi (1 - \cos \phi) + \frac{\sin^2 \phi (1 - \cos \phi)^2}{\dfrac{L_2^2}{r_5^{\,2}} - (1 - \cos \phi)^2} \right] \right\}$$

(2.28)

The motion of the output pinion:

$$\psi = \phi - \left(\frac{s_r}{r_5} \right) \left(\frac{180}{\pi} \right)$$

(2.29-1)

$$\dot{\psi} = \omega - \frac{v_r}{r_5}$$

(2.29-2)

$$\ddot{\psi} = -\frac{a_r}{r_5}$$

(2.29-3)

Observing the above equations, we see that both the output pinion's angular velocity $\dot{\psi}$ and angular acceleration $\ddot{\psi}$ are equal to zeros exactly when $\phi = 0^0, 360^0$; ϕ is measured counterclockwise from *Y*-axis as the baseline.

2.1.3 The Feasibility of Double Rocker Mechanism as a basic mechanism

Double rocker mechanism is rarely used unless another mechanism motivates it. A numeral example about a geared double rocker mechanism will be given in part IV Chapter 15.

2.14 Slider Crank Mechanism as a Basic Mechanism

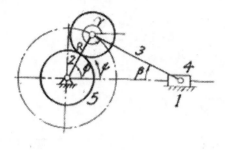

(a)

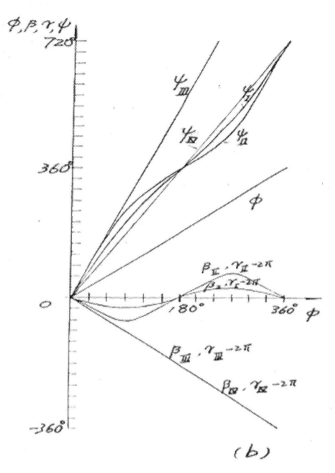

(b)

Figure 2.19   Two-gear four-bar linkage with slider crank mechanism as a basic mechanism: (a) configuration;(b)curves of angular displacement of coupler and output pinion versus that of input crank; $\beta_I, \beta_{II}, \beta_{III}$ and $\psi_I, \psi_{II}, \psi_{III}$ when $R/L = 1/2, 1/1.1, 1/1$ respectively. The gear ratio $n = -1$. Besides, β_{IV} and ψ_{IV} are presented when $R/L = 1/1, n = -1/2$

Fig. 2.19a shows a two-gear four-bar linkage with a slider crank mechanism as a basic mechanism. The number of degrees of freedom of this mechanism
F=3(5-1)-2×5-1×1=1

Let the lengths of crank and coupler be R and L respectively. The crank rotates counterclockwisely with a uniform speed ω. Links2,3, and5 is an epicyclic gear train, we tabulate their rotating relationships as follows:

Table 2.1 Epicyclic Gear Train Relationships

	Arm 2	Link 3	Gear 5
Motion with arm Relative to frame	ϕ	ϕ	ϕ

Motion relative to Arm	0	$-(\phi-\gamma)$	$+\dfrac{r_3}{r_5}(\phi-\gamma)$
Total motion relative to frame	ϕ	γ	$\phi+\dfrac{r_3}{r_5}(\phi-\gamma)$

Fig.2.19b shows curves of angular displacement of coupler and output pinion when different R/L and n occur. We find that gear ratio n (i.e. r_3/r_5) has great influence on the output ψ ,whereas the ratio R/L does on the coupler's γ .

The following restriction should be obeyed,i.e.

$$0 \le \frac{R}{L} \le 1$$

Suppose L is as long as infinity while R is constant,then a Scotch-yoke is present. If $L=R$, the slider may rush through the change point due to inertia which raises the stroke up to $4R$ while γ is continuing to rotate in the opposite sense of crank rotation. Whereas from Table2.1,we get

$$\psi = \phi + \frac{r_3}{r_5}(\phi-\gamma)$$

(2.30)

It should be mentioned that all the angles are having the subscripts oj's in the equation actually.

When $\dfrac{r_3}{r_5}=1, \phi=2\pi,$ then

$$\gamma = -2\pi$$

$$\psi_{II} = 2\pi + [2\pi - (-2\pi)] = 6\pi$$

This ψ_{III} is a straight line when $n= -1, R=L$; ψ_{IV} is another straight line when $n= -1/2, R=L$. And when $\phi = 2\pi,$ then

$$\psi_{IV} = 2\pi + \frac{1}{2}[2\pi - (-2\pi)] = 4\pi,$$ see Fig.2.19b.

Differentiating $\psi = \phi + \dfrac{r_3}{r_5}(\phi-\gamma)$ with respect to time once and twice, we have

$$\dot{\psi} = \omega + \frac{r_3}{r_5}(\omega - \dot{\gamma})$$

(2.31)

$$\ddot{\psi} = -\frac{r_3}{r_5}\ddot{\gamma}$$

(2.32)

So the relation between γ and ϕ will be the key.

From Fig.2.19a,we have

$$R\sin\phi = -Ls\sin\gamma, \quad \text{then}$$

$$\sin\gamma = -\frac{R}{L}\sin\phi$$

(2.33)

or $\qquad \sin\beta = -\dfrac{R}{L}\sin\phi$, because $\gamma = \beta + 2\pi$

Differentiating it with respect to time , yields

$$\frac{\dot{\gamma}}{\omega} = -\frac{R}{L}\left(\frac{\cos\phi}{\cos\gamma}\right) = -\frac{R}{L}\left[\frac{\cos\phi}{\left(1 - \dfrac{R^2}{L^2}\sin^2\phi\right)^{\frac{1}{2}}}\right]$$

(2.34)

further, we have

$$\frac{\ddot{\gamma}}{\omega^2} = -\frac{R}{L}\left[\frac{\sin\phi\left(\dfrac{R^2}{L^2} - 1\right)}{\left(1 - \dfrac{R^2}{L^2}\sin^2\phi\right)^{\frac{1}{2}}}\right]$$

(2.35)

Then a numerical example is presented for explaination.

[Example 2.2] Let the slider crank mechanism based two-gear four-bar mechanism has L=150 mm,R=50 mm,$r_3 = r_5 == 25$ mm,input crank's constant angular velocity $\omega = 500$ 1/s,when $\phi = 60^0$.Find the output angular velocity $\dot{\psi}$.

Solution:

$$\frac{R}{L} = \frac{50}{150} = \frac{1}{3}$$

Substituting the related values to eq.(2.34),we have

$$\frac{\dot{\gamma}}{\omega} = -\frac{1}{3} \left[\frac{\cos 60^0}{\left(1 - \frac{1}{3^2} \sin^2 60^0\right)^{\frac{1}{2}}} \right] = -0.174$$

$$\dot{\gamma} = -0.174\left(500^2\right) = -87 \quad 1/s$$

By eq.(2.31) yields

$$\dot{\psi} = 500 + \frac{25}{25}(500 + 87) = 1087 \quad 1/s$$

Observing the curves of ψ_I, ψ_{II} in Fig.2.19b, we find that the maximun and minimun inclination are at $\varphi = 0^0, 180^0$ i.e.the extremes of output angular velocities exist.

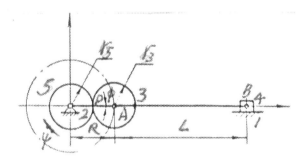

Figure 2.20   When $\phi = 0^0, \dot{\psi} = \dot{\psi}_{max}$

Fig. 2.20 shows that when $\phi = 0^0$, the output pinion's angular velocity $\dot{\psi}$ will be maximum. The proof is as follows:

Let

$$L \rangle R, r_3 = r_5, r_3 + r_5 = R$$
$$V_A = R\omega$$

The velocity of the pitch point P is

$$V_P = r_5 \dot{\psi} :$$

At this instant, the absolute instantaneous center of link 3 is on point B, therefore

$$\frac{V_A}{V_P} = \frac{L}{L + r_3}$$

$$\dot{\psi} = \frac{1}{r_5}\left(\frac{L + r_3}{L}\right)V_A = \frac{(L + r_3)R\omega}{L(R - r_3)} = \dot{\psi}_{max}$$

(2.36)

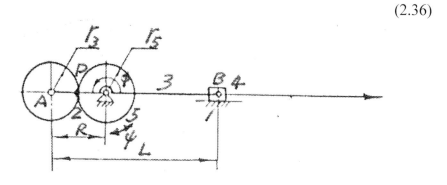

Figure   2.21   When $\phi = 180^0, \dot{\psi} = \dot{\psi}_{min}$

Fig.2.21 shows that when $\phi = 180^0$, $\dot{\psi}$ will be $\dot{\psi}_{min}$.   The proof is as follows:

$$\frac{V_A}{V_P} = \frac{L}{L - r_3} \qquad \dot{\psi} = \frac{1}{r_5}\left(\frac{L - r_3}{L}\right)V_A = \frac{(L - r_3)}{L(R - r_3)}R\omega = \dot{\psi}_{min}$$

(2.37)

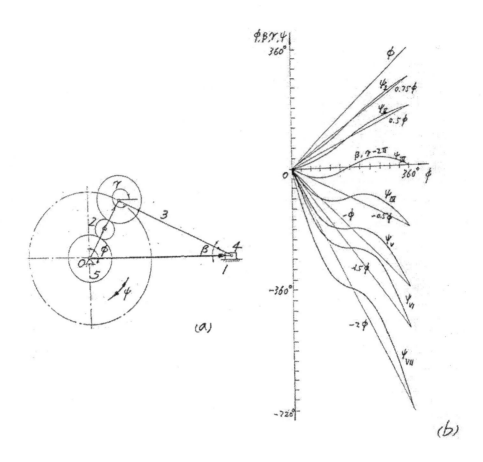

Figure 2.22　　Mechanism with an added idler gear： (a)configuration;(b) output angular displacement curves when $R/L = 1/2, n = 0.25 \sim 3.0$

It may get some special effect when an idler gear is added as in Fig. 2.22a. The relations of the angular displacement among menbers 2,3,and 5(between n and ψ) will be shown in Table 2.2 and 2.3 respectively

Table　　2.2

	Arm 2	Link 3	Gear 5
Motion with arm relative to frame		ϕ	ϕ
Motion relative to arm	0	$-(\phi - \gamma)$	$-\dfrac{r_3}{r_5}(\phi - \gamma)$
Total motion relative to　frame	ϕ	γ	$\phi - \dfrac{r_3}{r_5}(\phi - \gamma)$

If $r_3 = r_5$, then the output angular displacement ψ equals to $\phi - (\phi - \gamma) = \gamma$. That is, the motion of gear 5 is exactly the same as that of link 3. Gear 5 oscillates as link 3 does. Table 2.3 shows the relationship of ψ with different n values if $R/L = 1/2$. Fig.2.22b shows the curves

Table 2.3

	n	ψ
I	0.25	$0.75\phi + 0.25\gamma$
II	0.5	$0.5\phi + 0.5\gamma$
III	1	γ
IV	1.5	$-0.5\phi + 1.5\gamma$
V	2	$-\phi + 2\gamma$
VI	2.5	$-1.5\phi + 2.5\gamma$
VII	3	$-2\phi + 3\gamma$

Note:　When $R/L=1/2$

of the corresponding values of ψ, γ. Look at these curves, we find that if $r_3 \geq r_5$, then the output gear reverses partly or wholly instead of forward. And there are extremes existing in some curves, they can be the clues of dwell.

If an internal gear replaces gear 5 (Fig.2.23), the output motion of gear 5 is the

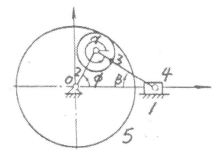

Figure　2.23　　Mechanism with internal gear

same as the external gearing with an idler gear.　However, r_3 must be smaller than r_5 due to restriction of construction. Curves on Fig.2.22b can be referred only on the upper half. The extremes of the output angular velocity will be derived bellow

78

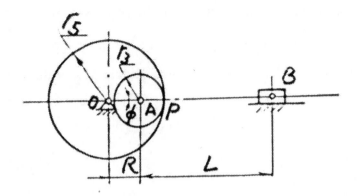

Figure 2.24 When $\phi = 0^0, \dot{\psi} = \dot{\psi}_{min}$. Links not shown.

First, when $\phi = 0^0$, Fig.2.24

$$\frac{V_A}{V_P} = \frac{L}{L - r_3}$$

$$\dot{\psi} = \frac{V_P}{r_5} = \frac{1}{r_5}\left(\frac{L - r_3}{L}\right)V_A = \frac{(L - r_3)R\omega}{L(R + r_3)} = \dot{\psi}_{min}$$

(2.38)

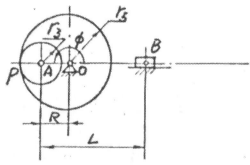

Figure 2.25 When $\phi = 180^0, \dot{\psi} = \dot{\psi}_{max}$. Links not shown.

Secondly, when $\phi = 180^0$,Fig.2.25.

$$\frac{V_A}{V_P} = \frac{L}{L + r_3}$$

$$\dot{\psi} = \frac{1}{r_5}\left(\frac{L + r_3}{L}\right)V_A = \frac{(L + r_3)R\omega}{L(R + r_3)} = \dot{\psi}_{max}$$

(2.39)

Furthermore, if link 3 becomes an internal gear (Fig.2.26), what will happen?   As

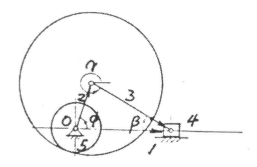

Figure   2.26   Link 3 becomes an internal gear

Table 2.4   Internal Gear Relationships

	Arm 2	Link 3	Gear 5
Motion with arm relative to frame	ϕ	ϕ	ϕ
Motion relative to arm	0	$-(\phi-\gamma)$	$\dfrac{r_3}{r_5}(\phi-\gamma)$
Total motion relative to frame	ϕ	γ	$\phi-\dfrac{r_3}{r_5}(\phi-\gamma)$

we   see from table 2.4, it is the same as the mechanism with an idler gear.   However, the ratio $r_3\big/r_5$ must be greater than unity whereas the curves referred should be the lower half of Fig.2.22b. The proof   of extremes will be:

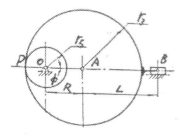

Figure   2.27   When $\phi=0^0, \dot{\psi}=\dot{\psi}_{max}$. Links not shown

When $\phi=0^0$, Fig.2.27

$$\frac{V_A}{V_P}=\frac{L}{L+r_2+r_5}=\frac{L}{L+r_3}$$

$$\dot{\psi} = \frac{V_P}{r_5} = \left(\frac{L+r_3}{Lr_5} \right) R\omega = \dot{\psi}_{max}$$

$$(2.40)$$

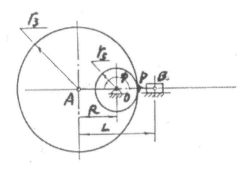

Figure 2.28 When $\phi = 180^0, \dot{\psi} = \dot{\psi}_{min}$. Links not shown.

When $\phi = 180^0$, Figure 2.28

$$\frac{V_A}{V_P} = \frac{L}{L-R-r_5} = \frac{L}{L-r_3}$$

$$\dot{\psi} = \frac{V_P}{r_5} = \left(\frac{L-r_3}{Lr_5} \right) R\omega = \left(\frac{L-R-r_5}{Lr_5} \right) R\omega = \dot{\psi}_{min}$$

$$(2.41)$$

And when $L-R=r_5$, $\dot{\psi}$ would be zero. Besides, it can be proved by the coincidence of B (i.e.instantaneous center 13) — the intersecting point of "frame" OA and "coupler" 4 — and pitch point(i.e. instantaneous center 35) of the two gears,see Fig.2.29a II. Furtheremore, $\ddot{\psi} = 0$ because the collineation axis(i.e.OB at this moment)is perpendicular to the "coupler" 4 which has a infinite long length lying perpendicular to its stroke. We also can prove that by Fig.2.29b, curve $\dot{\psi}_{II}$'s slope once is zero (i.e.$\ddot{\psi}$ is zero)at the 180^0 Of ϕ where $\dot{\psi}_{II}$ is zero. The slope of curve $\dot{\psi}_{II}$ is zero at another point 0^0 of ϕ where $\dot{\psi}_{II} = -2\omega$, though the $\ddot{\psi}$ is zero.

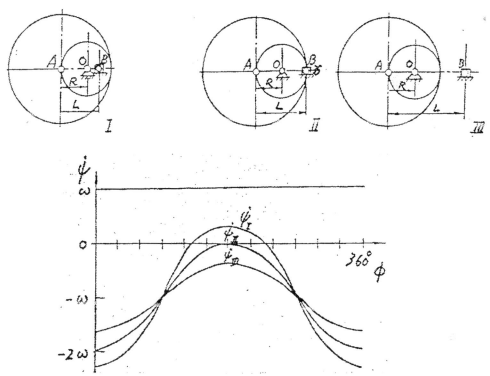

Figure 2.29 Three configurations of the mechanism (links not shown) (a) I, $(L-R)<r_5$;II, $L-R=r_5$;III, $(L-R)>r_5$;(b) curves of the output angular velocity $\dot{\psi}_I,\dot{\psi}_{II},\dot{\psi}_{III}$ with ω as its ordinate unit.

When $(L-R)>r_5$(e.g. R/L=1/3 in Fig.2.29aIII),gear 5 has no dwell, only rotating in varies speed, Fig.2.29b's curve $\dot{\psi}_{III}$ .When $(L-R)<r_5$ (e.g. R/L=1/1.5 in Fig.2.29aI),gear 5 has two first order dwells, i.e. angular velocity $\dot{\psi}$ is zero, reverse motion in the interval, also see Fig. 2.29b's curve $\dot{\psi}_I$.

Since reverse motion of the output member is important in dwell, we should take a look at it. There are three cases of $(L-R)<r_5$,Fig. 2.30. We choose(c)$r_5<\dfrac{r_3}{2}$ as an example for discussing,Fig.2.31.

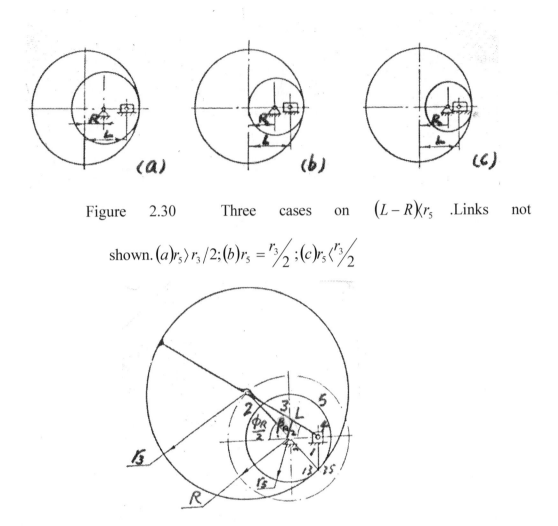

Figure 2.30 Three cases on $(L-R)\langle r_5$ .Links not shown. $(a)r_5\rangle r_3/2;(b)r_5=\dfrac{r_3}{2};(c)r_5\langle\dfrac{r_3}{2}$

Figure 2.31 $(L-R)\langle r_5,r_5\langle\dfrac{r_3}{2}$,one of the first order dwell position where $\dot\psi=0$.

As we all know, the coincidence of instantaneos centers 13 and 35 shows that the first order dwell is existent,i.e. $\dot\psi=0$. Fig.2.31 is exactly the configuration we need, and there would be another just located symmetrically about X-axis on the other side. Now let ϕ_R be the crank angular displacement during he reversing motion period , the angular displacemnt of link 3 will be β_R .The relationship between them should be

$$\psi_R=\left[\phi_R-\frac{r_3}{r_5}(\phi_R-\beta_R)\right]$$

They all subtend the arc in second and third quadrant separately,each equally divided at the either side of X-axis. And the half angle of ϕ_R and β_R will be discussed.

$$\left(R+r_5\right)^2 \cos^2\frac{\phi_R}{2} = L^2 \cos^2\frac{\beta_R}{2}$$

whereas $\qquad R^2 \sin^2\frac{\phi_R}{2} = L^2 \sin^2\frac{\beta_R}{2}$

therefore $\qquad \left(R+r_5\right)^2 \cos^2\frac{\phi_R}{2} = L^2\left(1-\frac{R^2}{L^2}\sin^2\frac{\phi_R}{2}\right)$

$$\cos\frac{\phi_R}{2} = \left[\frac{L^2-R^2}{r_5\left(r_5+2R\right)}\right]^{\frac{1}{2}}$$

$$(2.42)$$

[Example 2.3] Let the above mechanism has $R=20\text{mm}, r_5=20\text{mm}, L=30\text{mm}$. How many degrees are ϕ_R, ψ_R?

Solution:

$$r_3 = R+r_5 = 40\,\text{mm}$$

$$L-R = 30-20 = 10 \langle r_5 = 20$$

So the output gear has one reversing period and first order dwell twice.

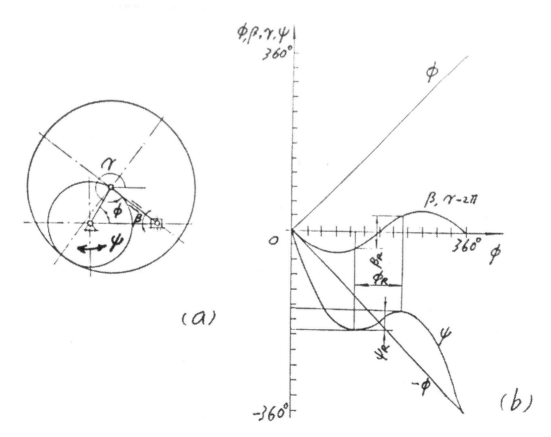

(a)

(b)

Figure 2.32 Mechanism from example 2.3, $(L-R)(r_5, r_5 = \frac{r_3}{2}$: (a) configuration;(b)curves of β, ψ

$$\cos\frac{\phi_R}{2} = \left[\frac{L^2 - R^2}{r_5(r_5 + 2R)}\right]^{\frac{1}{2}} = \left[\frac{30^2 - 20^2}{20(20 + 2 \times 20)}\right]^{\frac{1}{2}}$$

$$\frac{\phi_R}{2} = 49.793 0 3$$

$$\phi_R = 99.58668^0$$

$$\sin\frac{\beta_R}{2} = \frac{R}{L}\sin\frac{\phi_R}{2}$$

$$\frac{\beta_R}{2} = 33.987844^0$$

$$\psi_R = 2\left[49.793074^0 - 2\left(49.743034^0 - 33.987844^0\right)\right]$$

$$= 36.365228^0$$

Fig.2.32 diagrammatically shows the relationships among ϕ_R, β_R, ψ_R. Looking to it makes us understand it more clearly

If the two-gear four-bar linkage with offset slider crank mechanism is used as a basic mechanism, the characteristics of three cases — (i)nonstop,irreversible;(ii)irreversible,stop once only;(iii) reversing partly,stop twice ——will appear also. These are not to be discussed here.

2.1.5 Crank Oscillating Block Mechanism as a Basic Mechanism

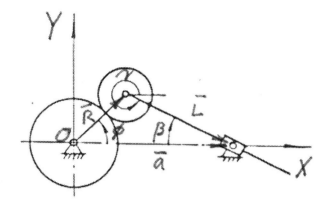

Figure   2.33   Two-gear four-bar linkage with crank oscillating block mechanism
as a basic mechanism

Fig. 2.33 shows a two-gear four-bar linkage with crank   oscillating block mechanism
as a basic mechanism,

its vector equation of closure is

$$\overline{R} + \overline{L} = \overline{a}$$

Their projections on Y-axis are

$$R\sin\phi = -L\sin\gamma$$

or $\qquad\qquad R\sin\phi = -L\sin\beta \text{,for } \gamma = \beta + 2\pi$

$$L = \left(R^2 + a^2 - 2aR\cos\phi\right)^{\frac{1}{2}}$$

$$\sin\gamma = \frac{-R\sin\phi}{\left(R^2 + a^2 - 2aR\cos\phi\right)^{\frac{1}{2}}}$$

$$\cos\gamma = \frac{\left(R^2 + a^2 - 2aR\cos\phi - R^2\sin^2\phi\right)^{\frac{1}{2}}}{\left(R^2 + a^2 - 2aR\cos\phi\right)^{\frac{1}{2}}}$$

$$= \frac{R\cos\phi - a}{\left(R^2 + a^2 - 2aR\cos\phi\right)^{\frac{1}{2}}}$$

$$(2.43)$$

Differentiating it with respect to time, we have

$$\frac{\dot{\gamma}}{\omega} = \frac{R(R - a\cos\phi)}{R^2 + a^2 - 2aR\cos\phi}$$

$$(2.44)$$

where ω is the constant angular velocity of the crank.

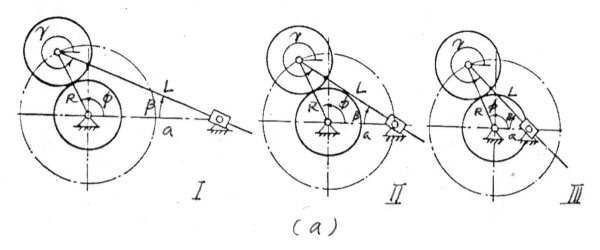

(a)

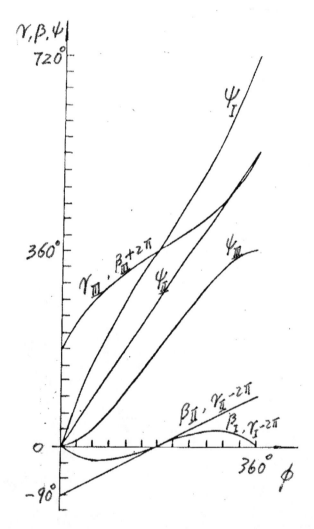

Figure 2.34 (a) Configurations $: n = -1; \dfrac{R}{a} = 0.5(I), 1(II), 2(III)$ respectively.(b)

Angular displacement curves of coupler and output gear: $\beta, \gamma; \psi$.

Differentiating it again with respect to time, we have

$$\frac{\ddot{\gamma}}{\omega^2} = \frac{aR\sin\phi\left(a^2 - R^2\right)}{\left(R^2 + a^2 - 2aR\cos\phi\right)^2}$$

(2.45)

From the equations above, we see that all the angular displacement ,angular velocity, and angular acceleration are the functions of ϕ, a, R. Furthermore, a and R can be combined to one parameter ,i.e. R/a, a may be defined as unity. Fig.2.34b shows the curves of angular displacement; Fig.2.35 shows the curves of specific angular velocity and acceleration of both coupler and output gear at various ratios of R to a.

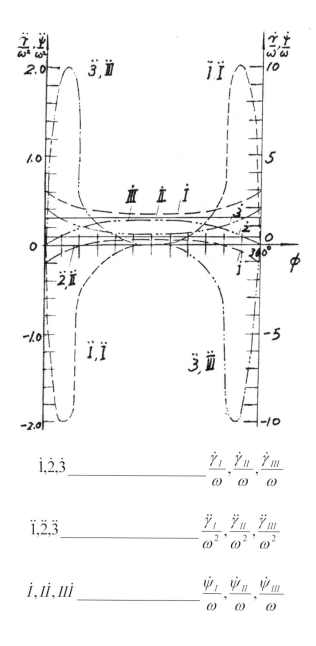

$\dot{1},\dot{2},\dot{3}$ _____ $\dfrac{\dot{\gamma}_I}{\omega}, \dfrac{\dot{\gamma}_{II}}{\omega}, \dfrac{\dot{\gamma}_{III}}{\omega}$

$\ddot{1},\ddot{2},\ddot{3}$ _____ $\dfrac{\ddot{\gamma}_I}{\omega^2}, \dfrac{\ddot{\gamma}_{II}}{\omega^2}, \dfrac{\ddot{\gamma}_{III}}{\omega^2}$

$\dot{I},\dot{II},\dot{III}$ _____ $\dfrac{\dot{\psi}_I}{\omega}, \dfrac{\dot{\psi}_{II}}{\omega}, \dfrac{\dot{\psi}_{III}}{\omega}$

$$\ddot{I}, \ddot{II}, \ddot{III} \underline{\hspace{4cm}} \frac{\ddot{\psi}_I}{\omega^2}, \frac{\ddot{\psi}_{II}}{\omega^2}, \frac{\ddot{\psi}_{III}}{\omega^2}$$

Figure 2.35 Curves of angular velocity to ω, acceleration to ω^2 ratios of both the coupler and output gear versus to crank rotating angles.

This drawing shows clearly what the behavior of the mechanism is, the puzzle of analytical method for the multiple solutions of inverse trigonometric function is to be avoided.

When $R/a=1$, curves of the angular displacement versus crank rotating angle of both the coupler and output gear are inclined straight lines (Fig.2.34); curves of angular velocity versus ω, horizontal lines (Fig.2.35); curves of angular acceleration versus ω^2, zero (Fig.2.35). And there is a phenomenon that the coupler revolves 180^0 while the input crank rotates 360^0, which will continue to complete its another 180^0 as the crank rotates the next 360^0.

Since $\gamma, \dfrac{\dot{\gamma}}{\omega}, \dfrac{\ddot{\gamma}}{\omega^2}$ may be calculated at any instant, however, the output gear's kinematic characteristics can be found by prescribed relationships of ϕ, γ, ψ as follows:

Since $\quad \psi = \phi + \dfrac{r_3}{r_5}(\phi - \gamma)$

Differentiating it with respect to time, we have

$$\dot{\psi} = \omega + \frac{r_3}{r_5}(\omega - \dot{\gamma})$$

or $\qquad \dfrac{\dot{\psi}}{\omega} = \dfrac{r_3}{r_5}\left(1 - \dfrac{\dot{\gamma}}{\omega}\right) + 1$

(2.46)

Furthermore,

$$\ddot{\psi} = -\frac{r_3}{r_5}\ddot{\gamma}$$

or

$$\frac{\ddot{\psi}}{\omega^2} = -\frac{r_3}{r_5}\left(\frac{\ddot{\gamma}}{\omega^2}\right)$$

(2.47)

These equations are just the same as that of the mechanism with slider crank mechanism as a basic mechanism.

2.1.6 Cycloidal Crank Mechanisms

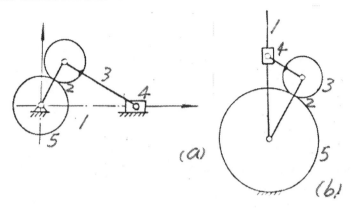

Figure 2.36   How the cycloidal crank mechanism comes from: (a) two-gear four-bar linkage ;(b) cycloidal crank mechanism.

If gear 5 (Fig.2.36a) is to be the frame instead of the output member of the two-gear four-bar linkage, then a cycloidal crank mechanism appears (Fig.2.36b). The reasons why this cycloidal crank mechanism is being so called are:first, link 3 can make complete turns relative to link 2 like a "crank" does , its pin connected to link 4 works as a "crankpin". Secondly, this crank may trace a cycloidal path. Besides, this mechanism is more widely used and studied than the cycloidal crank type two-gear five-bar linkage.

2.1.6.1   How the Epicycloidal Crank Mechanism is Generated and What
                 Motion It Has

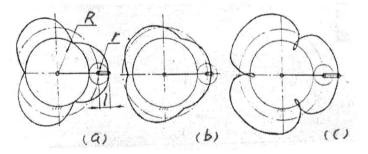

Figure 2.37   Three types of epicycloids, $R=3r$: $(a)l = r$;$(b)l \langle r$;$(c)l \rangle r$; where R—

—radius of sun gear, r——radius of planet gear, l——length of cycloidal crank.

Cycloidal crank is rigidly attached to planet gear. Strictly speaking, traces in (b),(c) may be called epitrochoid for $l \neq r$.

When the planet gear rotates around the sun gear (i.e. the central gear), the points on the planet gear will generate epicycloids which may have various shapes according to the ratio of the radius of sun gear to the radius of planet gear, R to r, and the ratio of the length of cycloidal crank to the radius of planet gear, l to r. Fig.2.37 shows the relationships of these parameters. There are three strands of curves as R is equal to $3r$.Fig.2.38 shows the mechanism of oblique drawing when $l \rangle r$. Fig.2.39 is the geometrical relationship between the input link 2 and output link 1. Now we are going to study the kinematic characteristics of the output link 1 and the extreme values located.

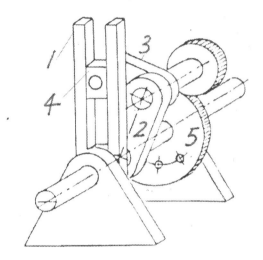

Figure 2.38   Oblique drawing of an epicycloidal crank mechanism

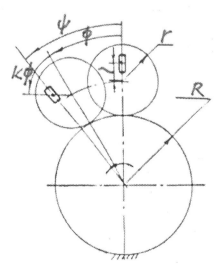

Figure 2.39   Geometry of an epicycloidal crank mechanism with $l = \dfrac{r}{2}$

Let $\quad K = \dfrac{R}{r}, N = \dfrac{l}{R+r}$, then the angular displacement, and its velocity, angular acceleration relevant to the input link's ω, ω^2 will express as follows:

$$\psi = \tan^{-1} \frac{\sin\phi + N\sin(K+1)\phi}{\cos\phi + N\cos(K+1)\phi}$$

(2.48)

$$\frac{d\psi}{\omega dt} = \frac{1 + (K+1)N^2 + (2+K)N\cos K\phi}{1 + N^2 + 2N\cos K\phi}$$

(2.49)

$$\frac{d^2\psi}{\omega^2 dt^2} = \frac{\left(-1+N^2\right)K^2 N\sin K\phi}{\left(1+N^2+2N\cos K\phi\right)^2}$$

(2.50)

The extreme positions of $\psi, \dot\psi, \ddot\psi$ in the crank angle positions are

$$\phi_{\psi extrema} = \frac{1}{K}\cos^{-1}\frac{-1-(K+1)N^2}{(2+K)N}$$

(2.51)

$$\phi_{\dot\psi extrema} = \frac{1}{K}\sin^{-1} 0^0$$

(2.52)

$$\phi_{\ddot\psi extrema} = \frac{1}{K}\cos^{-1}\frac{(1+N^2)\pm\sqrt{1+34N^2+N^4}}{4N}$$

(2.53)

The kinematic characteristics of a epicycloidal crank mechanism are shown in Fig.2.40. We see that the period of the complete motion cycle is 120^0 crank angle for the diameter of the planet gear is one-third of that of the sun gear. The three cases as non-stop, stop once, and stop twice that we have already met in some types of mechanisms will occur in this mechanism too. There is a reversing motion begins in about the first 49^0 crank angle position every motion cycle for $l = 1.5r$. Table 2.5 will show us the crank angle positions of the outstanding features.

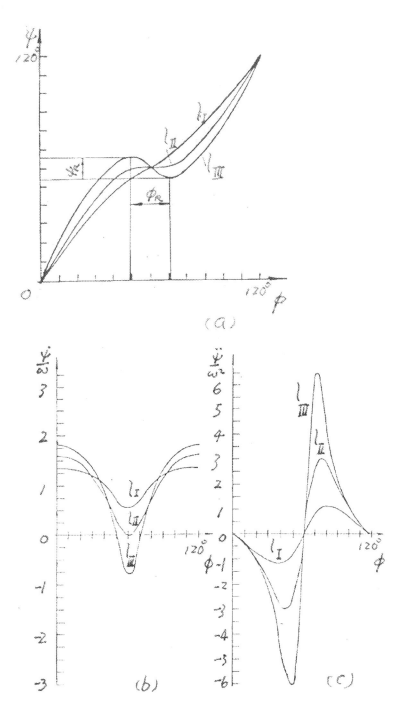

Figure 2.40 The Kinematic characteristics of the epicycloidal crank mechanism. Where $l_I = 0.5r$, $l_{II} = r$, $l_{III} = 1.5r$; $R:r = 3:1$;(a) angular displacement ψ ; (b) specific angular velocity $\dot{\psi}/\omega$;(c) specific angular acceleration $\ddot{\psi}/\omega^2$.

Table 2.5 Crank Angle Positions of the Outstanding Features (Epicycloidal)

$N=1/R+r$	$0.5/4$	$1/4$	$1.5/4$
$\phi_{\psi extema}$			48.81432^0 71.18577^0
ϕ_R			22.37154^0
$\psi_{extrema}$			65.592676^0 54.407302^0
ψ_R			11.184360^0
$\phi_{\dot{\psi}extrema}$		$0^0,60^0,120^0$	
$\dot{\psi}_{extrema}$	1.3313333 0.5714285	1.6 0	1.8127684 -0.8886435
$\phi_{\ddot{\psi}extrema}$	38.782644^0 81.217356^0	44.98019^0 75.01983^0	49.233668^0 70.766332^0
$\ddot{\psi}_{extrema}$	-1.2125424 1.2125424	-2.967677 2.967677	-6.0369692 6.0369692

2.1.6.2 Hypocycloidal Crank Mechanism Kinematic Characteristics

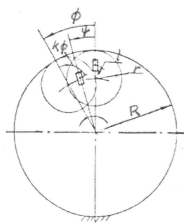

Figure 2.41   Geometry of a hypocycloidal crank mechanism with $l=r/2$

The displacement equation of a hypocycloidal crank mechanism is much alike to that of a epicycloidal's.Differences exist in some"+"and"-"signs only.

Let $K=\dfrac{R}{r}, M=\dfrac{l}{R-r}$, then we have

$$\psi = \tan^{-1}\frac{\sin\phi - M\sin(K-1)\phi}{\cos\phi + M\cos(K-1)\phi}$$

$$(2.54)$$

$$\frac{d\psi}{\omega dt} = \frac{1 - (K-1)M^2 + (2-K)M\cos K\phi}{1 + M^2 + 2M\cos K\phi}$$

(2.55)

$$\frac{d^2\psi}{\omega^2 dt^2} = \frac{(1-M^2)MK^2 \sin\phi}{(1 + M^2 + 2M\cos K\phi)^2}$$

(2.56)

$$\phi_{\psi extrema} = \frac{1}{K}\cos^{-1}\frac{-1 + (K-1)M^2}{(2-K)M}$$

(2.57)

$$\phi_{\dot\psi extrema} = \frac{1}{K}\sin^{-1}0^0$$

(2.58)

$$\phi_{\ddot\psi extrema} = \frac{1}{K}\cos^{-1}\left[\frac{(1+M^2)\pm\sqrt{1 + 34M^2 + M^4}}{4M}\right]$$

(2.59)

The kinematic characteristics of a hypocycloidal crank mechanism are shown in Fig. 2.42. Comparing it with the epicycloid's, we see that some are similar to each other, whereas some have differences between them The following table will show the characteristics in the crank angle positions and the others.

Table 2.6 Crank Angle Positions of the Outstanding Features (Hypocycloidal)

$M = l/R-r$	$0.5/2$	$1/2$	$1.5/2$
$\phi_{\psi extrema}$			33.1980230^0 86.801977^0
ϕ_R			66.396046^0
$\psi_{extrema}$			-7.0049436 127.004704^0
ψ_R			14.0096476^0
$\phi_{\dot\psi extrema}$	$0^0, 60^0, 120^0$		
$\dot\psi_{extrema}$	0.4 2.0	0.0 4.0	-0.2857142 10.0
$\phi_{\ddot\psi extrema}$	44.98115^0 75.01995^0	52.365647^0 67.634353^0	56.82796^0 63.17204^0
$\ddot\psi_{extrema}$	2.9676436 -2.9676436	12.09814 -12.09814	70.629628 -70.629628

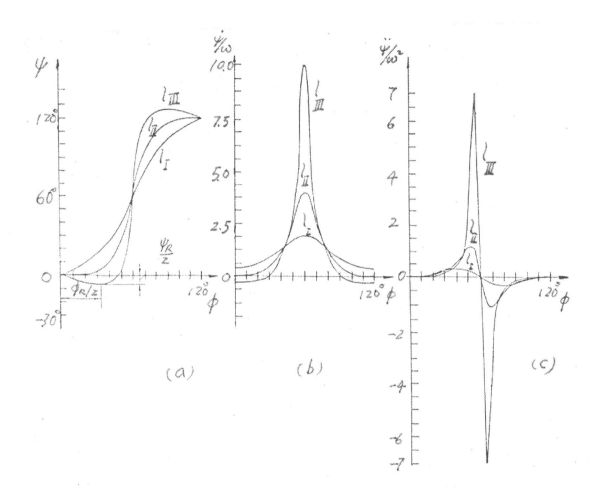

Figure 2.42   Kinematic characteristics of hypocycloidal crank mechanism.
Where $l_I = 0.5r, l_{II} = r, l_{III} = 1.5r,; R : r = 3 : 1$. (a) angular displacement ψ ; (b)

specific angular velocity $\dot{\psi}/\omega$; (c) specific angular acceleration $\ddot{\psi}/\omega^2$.

2.1..6.3   Applications of Cycloidal Crank Mechanism in Industry

The dwell in mechanism operation is very common. Fig.2.43 shows a mechanism
used for dwell. The ratio of the radius of the sun gear to that of planet gear
$R : r = 2 : 1$, the length of the cycloidal crank $l = 0.35r$. While the trace of the
crankpin deviates the horizontal line with $\pm 15^0$, the horizontal component of the
crankpin movement is about $0.0013r$, approaching to zero. This characteristics
makes the adjoining member dwelling in 30^0 interval of the driving member at
the two teminals of the output stroke approximately.

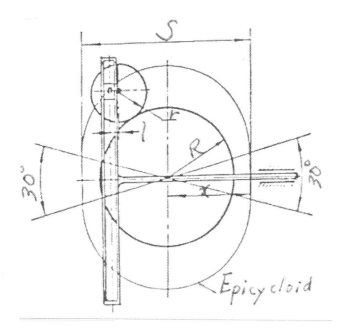

Figure 2.43 Example of epicycloidal crank mechanism, $R = 2r, l = 0.35r, s$ (length of the stroke) $= 5.3r$.

The above example is for epicycloidal crank mechanisms. Now we are going to see some for hypocycloidal crank mechanisms below.

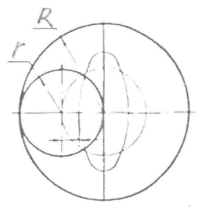

Figure 2.44 First example of hypocycloid crank mechanisms, $R = 2r, l = 0 \sim r$

Fig.2.44 is the example of a hypocycloidal crank mechanism with $R : r = 2 : 1$.. While l varies from $0 \sim r$, the path of crankpin will vary from a circle to a series of ellipses and finally a straight line across the internal gear being the length of the diameter.

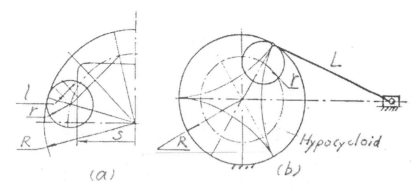

Figure 2.45　Another two examples of hypocycloidal crank mechanisms. $(a)R:r=4:1, l=0.402r, s=5.196r; (b)R:r=3:1.$

Figure2.45　is another two examples (a) shows a crankpin path being a square circumference with fillets(It dwells twice); (b) dwells once,the length of the coupler is equal to the mean radius of the"circular path"(the hypocycloid,in fact) of the crankpin approximately

2.2 .Three-Gear Four-Bar Linkage

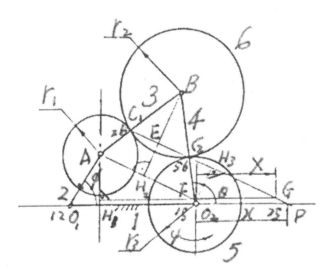

Figure 2.46　Three-gear four-bar linkage

The three-gear four-bar linkage shown in Fig.2.46 is the commonly adopted geared linkage in industry. Its characteristics have general significance, so studying it is worthwhile. The number of degrees of freedom of this linkage (Note that the gear 2 is rigidly attached to link 2.)

$$F = 3(6-1) - 2\times6 - 1\times2 = 1$$

The transmission ratio

$$n = \frac{\dot{\psi}}{\dot{\phi}} = \frac{x + L_1}{x}$$

(2.60)

where

$\dot{\psi}$ _____ output angular velocity of gear 5

$\dot{\phi}$ _____ input angular velocity of link 2

x _____ distance from instantaneous center 25 to the rotating center O_2 of gear 5

L_1 _____ length of link 1

From gear 2's rotating center A draw a line parallel to line 26-56 (i.e. C_1C_2) intersecting point F with line BO_2. Passing through point F draw a line 26-56 at G, intersecting the vertical line from A at H_1, we have

$$x = \frac{r_3}{r_1} X$$

(2.61)

where r_3, r_1 are the pitch circle radii of gears 5 and 2 respectively, X is the length of FG. Since

$$\Delta C_2 P O_2 \frown \Delta C_2 GF \text{ ,therefore}$$

$$BH_2 = \sqrt{L_3^2 - \left(\frac{L}{2}\right)^2} = \frac{1}{2}\sqrt{4L_3^2 - L^2}$$

where $L = AF$

Draw lines BH_2, O_2H_3 perpendicular to AF and C_1P (i.e.26-25) respectively, we have points E, H_2, H_3.

$$BE = \frac{r_2(BH_2)}{L_3}$$

$$FH_3 = EH_2 = BH_2 - BE = \frac{1}{2}\sqrt{4L_3^2 - L^2\left(\frac{r_3}{L_3}\right)}$$

$$AH_1 = L_2 \sin\phi + (r_1 - r_3)\sin\theta$$

$$\frac{X}{L} = \frac{FH_3}{AH_1}$$

$$X = \frac{Lr_1\sqrt{4L_3^2 - L^2}}{2L_3\left(L_2\sin\phi + m\sin\theta\right)} \tag{2.62}$$

where

$$m = r_1 - r_3$$

$$L^2 = L_1^2 + L_2^2 + m^2 + 2mL_2\cos(\phi-\theta) + 2L_1\left(L_2\cos\phi + m\cos\theta\right)$$

r_2, L_2, L_3, θ are the radius of gear 2, lengths of links 2 and 3, and the position

angle of link 4 respectively.

By eqs.(2.60),(2.61),and(2.62), we have

$$\psi = \int\left\{1 + \frac{2L_1L_3\left[L_2\sin\phi + (r_1-r_3)\sin\theta\right]}{Lr_3\sqrt{4L_3^2 - L^2}}\right\}d\phi \tag{2.63}$$

This is the angular displacement equation of gear 5 with respect to input angle ϕ.

2.3 Four-Gear Four-Bar Linkage

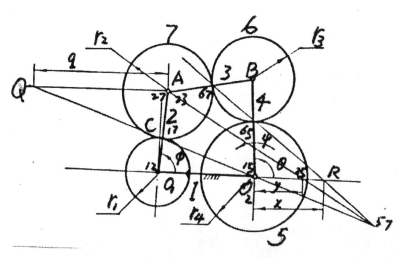

Figure 2.47   Four-gear four-bar linkage

The number of degrees of freedom of the four-gear four-bar linkage (Fig.2.47)

$$F = 3(7-1) - 2 \times 7 - 1 \times 3 = 1$$

With the help of instantaneous center 25, we have the transmission ratio

$$n = \frac{\dot{\psi}}{\dot{\phi}} = \frac{y + L_1}{y}$$

(2.64)

where $y = 15\text{-}25$

Passing instantaneous center 23 (i.e. A) draw a line parallel to link 1 intersecting line 15-17 (i.e. O_2C) at point Q. Hence

$$\triangle AQC \sim \triangle O_1O_2C$$

therefore

$$q = \frac{r_2}{r_1} L_1$$

(2.65)

$$y = \frac{r_4 x L_1}{r_1 x + r_4 L_1}$$

(2.66)

where

$$x = RO_2$$

Substitute eq.(2.66) into eq.(2.64) then

$$\dot{\psi} = \left(1 + \frac{L_1}{x} + \frac{r_1}{r_4}\right)\dot{\phi}$$

(2.67)

Substituting x into eq.(2.67) and integrating it with respect to input angle ϕ, we get the angular displacement equation

$$\psi = \int \left\{ 1 + \frac{2L_1L_3\left[L_2 \sin\phi + (r_2 - r_4)\sin\theta\right]}{Lr_4\sqrt{4L_3^2 - L^2}} + \frac{r_1}{r_4} \right\} d\phi$$

(2.68)

where L is the same as in paragraph 2.2.

2.4 .Remarks on Geared Four-Bar Linkages

(1) The two-gear four-bar linkage is the foundation of geared linkage with four bars. If we study this mechanism minutely, mechanisms with more gears will be

easy to investigate. No matter how many gears there are, the independent variable is only one of the gears; the others are restricted by corresponding link lengths, that is, the parameters of these gears are all dependent variables.

(2) Though the output functions are simply three cases, i.e.(i) non-stop, (ii) stop once, and (iii) stop twice; the joining of three or four gears instead of two must have had distinct changes which is worth studying.

(3) The critical (i.e. the output of a second order dwell does occur.) frame length of a three-gear four-bar linkage is known to academic circle, so it is neglected here.

3   Geared Six-Bar Linkage

3.1 Four-Gear Six-Bar Linkage

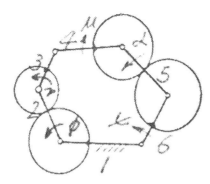

Figure 3.1   Four-gear six-bar linkage

In Fig.3.1, four gears are rigidly attached to corresponding links respectively in the four-gear six-bar linkage. Its degrees of freedom number

$$F = 3(6-1) - 2 \times 6 - 1 \times 2 = 1$$

The vector equation of closure

$$\overline{L}_{2j} + \overline{L}_{3j} + \overline{L}_{4j} + \overline{L}_{5j} + \overline{L}_{6j} = \overline{L}_{1j} = \overline{L}_{11} = 1$$

or

$$e^{i\phi_{1j}}\overline{L}_{21} + e^{i\gamma_{1j}}\overline{L}_{31} + e^{i\mu_{1j}}\overline{L}_{41} + e^{i\alpha_{1j}}\overline{L}_{51} + e^{i\psi_{1j}}\overline{L}_{61} = \overline{L}_{11} = 1$$

where $\phi_{1j}, \gamma_{1j}, \mu_{1j}, \alpha_{1j} \psi_{1j}$ are the angular displacements of links 2,3,4,5, and 6

from initial to jth position respectively.

Since there are two pairs of meshing gears, so that two relationships occur. The following two tables will demonstrate the two planetary gear trains in detail.

Table 3.1   Planetary Gear Train Relationship I

	Arm 2	Gear 1	Gear 3
Motion with arm relative to frame	ϕ	ϕ	ϕ
Motion relative to arm	0	$-\phi$	$\dfrac{N_1}{N_3}\phi$
Total motion relative to frame	ϕ	0	$\left(1 + \dfrac{N_1}{N_3}\right)\phi$

Table 3.2   Planetary Gear Train Relationship II

	Arm 5	Gear 6	Gear 4
Motion with arm relative to frame	α	α	α
Motion relative to arm	0	$\psi-\alpha$	$-(\psi-\alpha)\dfrac{N_6}{N_4}$
Total motion relative to frame	α	ψ	$\alpha-(\psi-\alpha)\dfrac{N_6}{N_4}$

If this linkage is used as a function generator, the function $\psi=f(\phi)$ is known, the angular displacements γ,μ can be calculated by the following equations with the arbitrarily chosen and prescribed parameters $\alpha, N_1, N_3, N_4, N_6$.

$$\gamma=\phi\left(1+\frac{N_1}{N_3}\right)$$

$$\mu=\alpha-(\psi-\alpha)\frac{N_6}{N_4}$$

The number of precision positions and other parameter's relationships are tabulated as follows:

Table 3.3　The Four-Gear Six-Bar Linkage Function Generator

No.of Precision positions	No. of Algebraic equations	Unknowns and their numbers	Arbitrarily chosen reals and their numbers	No. of unknowns to be solved for
$\Delta\phi,\Delta\psi,n_1 n_2$ prescribed; linear solution				
1	2	$\overline{L}_{21},\overline{L}_{31},\overline{L}_{41},\overline{L}_{51}\overline{L}_{61}$ 10	4 vectors of the left 8	2
2	4	Above$+\alpha_{12}$ 11	3 vectors of the left$+\alpha_{12}$ 7	4
3	6	Above$+\alpha_{13}$ 12	2 vectors of the left$+\alpha_{12},\alpha_{13}$ 6	6
4	8	Above$+\alpha_{14}$ 13	1 vector of the left$+\alpha_{12},\alpha_{13}\alpha_{14}$ 5	8
5	10	Above$+\alpha_{15}$ 14	$\alpha_{12}\alpha_{13}\alpha_{14}\alpha_{15}$ 4	10

$\Delta\phi, \Delta\psi, n_1, n_2$ prescribed; nonlinear solution				
6	12	Above+α_{16} 15	$\alpha_{12}, \alpha_{13}, \alpha_{14}$ 3	12
7	14	Above+α_{17} 16	α_{12}, α_{13} 2	14
8	16	Above+α_{18} 17	α_{12} 1	16
9	18	Above+α_{19} 18	0	18
Nothing prescribed; nonlinear solution				
10	20	Above+ $\alpha_{110}, \Delta\phi, \Delta\psi,$ n_1, n_2 23	α_{12}, n_1, n_2 3	20
11	22	Above+α_{111} 24	$n_1 n_2$ 2	22
12	24	Above+$\alpha_{1'12}$ 25	n_1 1	24
13	26	Above+α_{113} 26	0	26

Observing the above table, we see that if the linear solution of the four-gear six-bar linkage function generator is claimed the number of precision positions can be up to five. This function is better than any other mechanisms which we have discussed before. Besides, the upmost number of precision positions can be thirteen. As the procedure of the synthesis of the linkage is similar to the previous examples stated, it is neglected here

4. Geared Three- Bar Linkages

4.1 Planetary Gear Train,Rack, and Scotch Yoke Composition

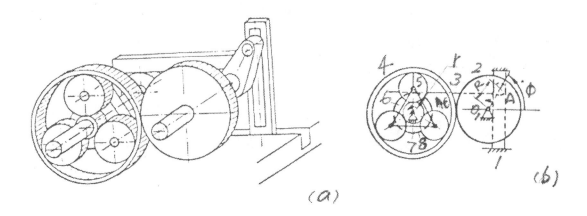

(a)

(b)

Figure 4.1   Planetary gear train, rack, and Scotch yoke composition: (a) oblique drawing; (b) configuration.

There are various planetary gear train linkages. Now we just take a typical one

——planetary gear train, rack, and Scotch yoke composition——as an example.

As a matter of fact, the mechanism shown in Fig.4.1 belongs to former multi-gear four- bar linkages. Pin A originally belongs to a hole on the slider, whereas now it composes a higher pair with the inner surface of the Scotch yoke instead of two lower pairs of the slider with yoke and the pin with hole (of course the slider is omitted). Though the output motion remains unchanged, the four-bar becomes three-bar. The input shaft is attached to crank and gear 2, on which the crankpin A forces the yoke making sinusoidal motion when gear 2 is meshing with the outer gear of the ring gear 4 rotating uniformly.

The number of degrees of freedom of the linkage

$$F = 3(7 - 1) - 2 \times 6 - 1 \times 5 = 1$$

For smooth operation, both the angular velocity and angular acceleration at the beginning and the end of the motion cycle of the output pinion should be zeros. Whereas the cycloidal projection motion (Fig.4.2) is one of the most common ones which can satisfy the requirement. That is

$$\psi = m_1 m_2 (\phi - \sin\phi)$$

(4.1)

where ψ is the x and $m_1 m_2$ is the R in the Figure.

And $m_1 = +\dfrac{N_2}{N_{4e}}, m_2 = +\dfrac{N_{4i}}{N_7}, N_2, N_{4e}$, N_{4i} , N_7 are the numbers of teeth of

gear2,external and internal gear 4, and gear 7 respectively. The ϕ is the angular displacement of the input gear 2.

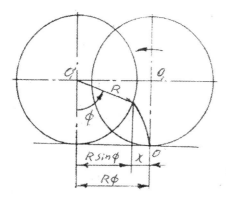

Figure 4.2   The corresponding cycloidal projection motion, $x = R(\phi - \sin\phi)$

The angular motion ψ of the output pinion 7 is the combination of the following two: one is $m_1 m_2 \phi$ which is completed by the action of uniform velocity motion of the input gear, the other is $m_1 m_2 \sin\phi$ which is completed by the yoke and the planetary gear train with two degrees of freedom. Table 4.1 shows the relationships of the members of the gear train,

Table 4.1   Planetary Gear Train Relationship III

	Arm 6	Gear 4	Pinion 7
Motion with arm relative to frame	θ	θ	θ
Motion relative to arm	0	X	$-m_2 X$
Total motion relative to frame	θ	$\theta + X$	$\theta - m_2 X$

Where θ is the angular displacement of spacer (arm) 6; $\theta + X$ is the same thing as $-m_1\phi$, the angular displacement of ring gear 4; $\theta - m_2 X$ is ψ, i.e. $m_1 m_2 (\phi - \sin\phi)$ By these relations, we have the angular displacement

$$\theta = \frac{-m_1 m_2 \sin\phi}{1 + m_2}$$

(4.2)

The displacement of rack 3

$$s = -\frac{m_1 m_2}{1 + m_2} r \sin\phi = -R \sin\phi$$

(4.3)

where r is the radius of the gear 8 which meshes with the rack, attached to the spacer 6; R is the length of the crank $O_1 A$. By eq.(4.3) yields

$$\frac{R}{r} = \frac{m_1 m_2}{1 + m_2}$$

(4.4)

Furthermore, we want to obtain the velocities and accelerations of other members by differentiation with respect to time.

$$\dot{\psi} = m_1 m_2 \omega \, (1 - \cos\phi)$$

(4.5)

$$\ddot{\psi} = m_1 m_2 \omega^2 \sin \phi$$

(4.6)

$$\dot{\theta} = -\frac{m_1 m_2}{1 + m_2} \omega \cos\phi$$

(4.7)

$$\ddot{\theta} = \frac{m_1 m_2}{1 + m_2} \omega^2 \sin \phi$$

(4.8)

$$\dot{s} = -\frac{m_1 m_2}{1 + m_2} r \omega \cos \phi$$

(4.9)

$$\ddot{s} = \frac{m_1 m_2}{1 + m_2} r \omega^2 \sin \phi$$

(4.10)

The curves in Fig.4.3 show the kinematic characteristics of the output pinion when $m_1 m_2 = 2$.

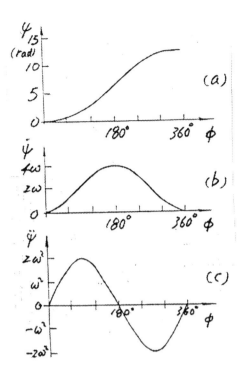

Figure 4.3   The kinematic charateristiccs of the output pinion when $m_1 m_2 = 2$:
(a)   angular displacement,(b) angular velocity;(c) angular acceleration.

Part II   Cam-Link Mechnisms

Introduction

No doubt, cam-link meanisms might be the finest composition in the industry if the wear of cam mechanism and the balance of mechanism were not to be considered. Generally speaking, cam-link mechanisms are able to get good results of particle trace, rigid body guidance, kinematic characteristics modification, and output amplitude expansion etc. However, considerations as how to reduce the wear and unbalance of members and how to modify the output functions are still the problems in reality. Research reports on these purposes have shown many noticeable results accomplished. The following cam-link mechanisms, Table II.1 from Kurt Hain, belong to 6 categories, 158 types.

Table II.1   Cam –Link Mechnisms

	Categories		Types	
1	Single Cam	Single joint	1	23
2		Double joints	24	44
3	Double Cam	Single joint	45	84
4		Double joints	85	122
5	Triple Cam	Single joint	123	140
6		Double joints	141	158

These mechanisms can be increased to more categories if the slider pairs replace pin joints.

Cam-link mechanisms come from the conventional six-bar linkage___Watt and Stephenson mechanisms.

Fig.II.1 shows ten inversions from Watt mechanism and thirteen from Stephenson's; total twenty-three.

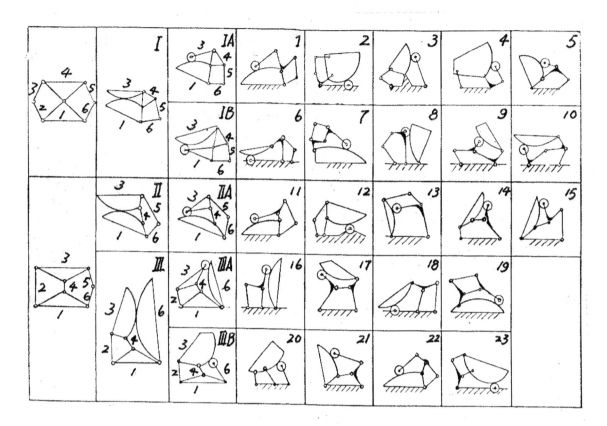

Figure II.1   Single cam single joint cam-link mechanisms

The Watt mechanism is more popular than the Stephenson's. We can easily see from Fig.II.1 the inversion of I,II,III that the cam pair instead of two revolute pairs with a link; either cam pair 1 and 3 instead of link 2 or cam pair 3 and 6 instead of link 5. Furthermore,one member of the matched cam pair becomes a roller follower, namely IA,IB,IIA,IIIA,andIIIB. In addtion, take different link to be frame by order, all twenty-three types are presented.

5   Cam-Link Mechanisms Connected in Series

5.1   Introduction

The cam-link mechanisms has many advantages: that is, the transmission angle
can satisfy the design requirements easily; the acceleration can be controlled in an
allowable range; and the output amplitude, for instance, the oscillating angle of an
oscillating member may raise up to a great extent. The cam-link mechanisms
connected in series might meet the above requirements satisfactorily, Fig.5.1.

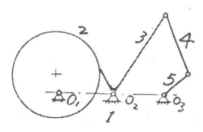

Figure 5.1   Cam-link mechanisms connected in series

The driving cam is assumed to be a constant rotating speed member. The follower
is the output member of the cam mechanism, the first component
mechanism ;simultaneously it is the input member of the four-bar linkage, the
second component mechanism.

Usually a double rocker mechanism is used as the second component mechanism
if the purposes of this cam-link mechanisms are: raising the amplitude of the
output oscillating angle and getting good kinematic characteristics, such as
shockless, smooth ,and reliable operation ,etc. No doubt, the extension of output
amplitude is more important, because the other kinematic characteristic
modifications can be easily fulfilled by the cam mechanism itself.

The number of degrees of freedom of the mechanism in Fig..5.1,

$$F = 3(5 - 1) - 2 \times 5 - 1 \times 1 = 1$$

If the required output oscillating angle Ψ_{II} of the oscillating link of the
second component (i.e. double rocker) mechanism is 150^0, the relevant
parameters will be presented by No.5 in Table I.1 of Appendix I They are: input
oscillating angle of second component mechanism

$$\Phi_{II} = 45^0 \sim 50^0 .(\text{corresponding to } \phi_1 \text{ of No.5}),$$

the allowable minimum pressure angle

$$[\mu_{min}]=45^0$$

the frame length

$$a_{II} = 1 \text{ (unity)},$$

and the ratios of other links

$$\frac{d_{II}}{a_{II}} = 1.5, \frac{f}{e} = \frac{\sqrt{2}}{2}, \frac{f}{a_{II}} = 0.5, \text{ all shown in Fig.5.2.}$$

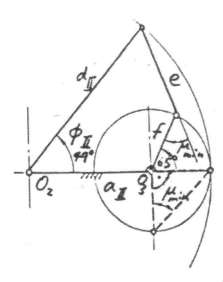

Figure 5.2 Double rocker mechanism (with large output oscillating angle) in two positions where the minimum pressure angles occur, one in its initial (solid) and the other when $\phi_{II} = 0$ (short dashed)positions.

The minimum pressure angle μ_{min} occurs twice in one motion cycle: one in the initial position(prescribed) and the other in the position when $\phi_{II} = 0^0$.

Figg.5.2 is the product of analog simulation,

$\Phi_{II} = 49^0, \Psi_{II} = 65^0 + 90^0 = 155^0$. If the output must be controlled exactly, i.e. 150^0 exactly no more no less; then some parameters should be changed. One way is that let the initial position of the output link intersects with the horizontal line forming an angle of 60^0 instead of 65^0 (Fig.5.2). Certainly, the μ will be a little more than μ_{min}, the Φ_{II} will be a little less than 49^0. These will do no harm to the mechanism, Fig.5.3.

Since the simple harmonic motion is good for operation, we choose it as the output motion. Let 150^0 be the amplitude of the output oscillating unity, i. e. $\Psi_{II} = 1$; $T=1$ sec., period unity; t partial T; ψ_{II}, corresponding angle displacement to t; then

$$\psi_{II} = \frac{\Psi_{II}}{2}\left(1 - \cos 2\pi \frac{t}{T}\right) = \frac{1}{2}\left(1 - \cos \frac{2\pi}{T}t\right) \times 150^0$$

(5.1)

$$\dot{\psi}_{II} = \frac{\pi}{T}\Psi_{II} \sin 2\pi \frac{t}{T} = \pi\left(\sin \frac{2\pi}{T}t\right) \times 150^0, 1/\sec.$$

(5.2)

$$\ddot{\psi}_{II} = \frac{2\pi^2}{T^2}\Psi_{II} \cos 2\pi \frac{t}{T} = 2\pi^2\left(\cos \frac{2\pi}{T}t\right) \times 150^0, 1/\sec^2.$$

(5.3)

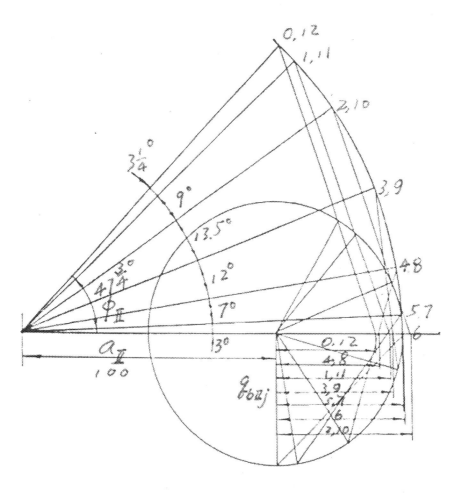

Figure 5.3 Twelve positions(with related parameter values) located according to the output simple harmonic motion, $a_{II} = 100$ mm.

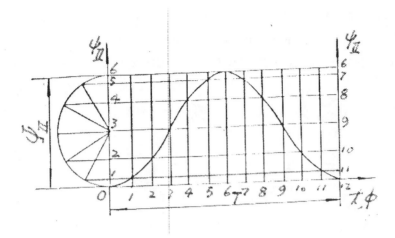

Figure 5.4   Simple harmonic motion

Fig.5.4 shows the curve of output angular displacement versus time.

5.2 Velocities and Accelerations of Mechanisms Connected in Series

It is necessary to know the velocities and accelerations of mechanisms connected in series for synthesis. Now let us take an equivalent four-bar linkage instead of the first component mechanism. Then the two component mechanisms are two four-bar linkages connected in series, Fig.5.5. The total transmission ratio

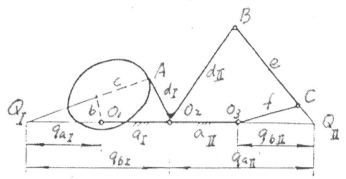

Fig 5.5   An equivalent four-bar linkage couples with another four-bar linkage in series.

of whole machine n_0 is equal to the product of the two component mechanisms transmission ratios $n_I \times n_{II}$. That is

$$n_0 = n_I \times n_{II}$$

(5.4)

where $n_I = \dfrac{\omega_{dI}}{\omega_b}, n_{II} = \dfrac{\omega_f}{\omega_{dII}}, \omega_{dI} = \omega_{dII} = \omega_d$;

then $n_0 = \dfrac{\omega_{dI}}{\omega_b} \times \dfrac{\omega_f}{\omega_{dII}} = \dfrac{\omega_f}{\omega_b}$,

(5.5)

where $\omega_b, \omega_{dI}, \omega_{dII}, \omega_f$ are the angular velocities of corresponding links

$b, d_I d_{II}, f$. Links d_I, d_{II} are just the two parts of the same link, so their angular velocities are the same.

Besides, by the instantaneous center theorem, we have

$$n_I = \frac{q_{aI}}{q_{bI}}, n_{II} = \frac{q_{aII}}{q_{bII}}$$

(5.6)

where $q_{aI}, q_{bI}, q_{aII}, q_{bII}$ are the distances among the instantaneous centers Q_I, Q_{II}

of corresponding mechanisms and the rotating centers O_1, O_2, O_3 of their driving

and driven shafts, see Fig.5.5.

The acceleration of a composite mechanism (the cam-link mechanism is a typical one) is expressed as "specific overall acceleration" (Appendix J)

$$A_0 = \frac{\varepsilon_f}{\omega_b^2} = \left[n_I^2 A_{II} + n_{II} A_I \right]$$

(5.7)

Where $A_I \left(= \frac{\varepsilon_d}{\omega_b^2} \right), A_{II} \left(= \frac{\varepsilon_f}{\omega_d^2} - \frac{\varepsilon_d}{\omega_b^2} \times \frac{\omega_f}{\omega_d} \right)$ are the "specific acceleration" of each

component mechanism; as shafts of b, d, f corresponding to shaft centers

O_1, O_2, O_3 from Appendix K, we have

$$\overline{A}_I = \frac{\overline{a}_I \times \overline{m}_d}{q_{bI}^2}, \overline{A}_{II} = \frac{\overline{a}_{II} \times \overline{m}_f}{q_{bII}^2}$$

where m_d, m_f correspond to m_I, m_{II}.

5.3 Maximum and Minimum Radii of the Cam

.It is obvious that the cam profile must be designed according to the motion of output member. The dimension of the cam should be as small as possible if only the allowable transmission angle is not exceed.

Let the cam shaft rotates at a uniform speed, and the leading half and the following half cycle are corresponding to a back and forth motion of output motion cycle. As the output motion cycle is divided into twelve equal parts, the relevant output angle ψ_{II} and corresponding input angle ϕ_{II} can be obtained either by analytical or by graphical method. The whole transmission ratio n_0 of

the cam-link mechanism is equal to ω_f / ω_b. On the other

hand, $n_0 = n_I n_{II}, n_{II} = q_{aII} / q_{bII}$; whereas q_{aII}, q_{bII} can be obtained by graphical

method, see Fig. 5.3.

Then

$$n_I = \frac{q_{aI}}{q_{bI}} = \frac{q_{bI} - a_I}{q_{bI}} = 1 - \frac{a_I}{q_{bI}}$$

$$q_{bI} = \frac{a_I}{1 - n_I}$$

(5.8)

The instantaneous center Q's of the input and output member of the equivalent mechanism will be obtained by corresponding q_{bI}'s. Twelve: Q's belonging to the twelve positions of one cycle will compose a closed loop later. Table 5.1 shows twelve q_{bI}'s and related data.

Table 5.1 Twelve q_{bI} and Related Data

No	t	ψ_{II}	ϕ_{II}	n_0	q_{bII}	n_{II}	n_I	q_{bI}
0	0	0^0	0^0	0	39.5	3.5316	0	100.0
1	$T/12$	10.048^0	3.25^0	0.6545	45.5	3.1978	0.2046	125.66
2	$T/6$	37.5^0	12.25^0	1.1336	53.0	2.8869	0.3925	167.7
3	$T/4$	75.0^0	25.75^0	1.3085	46.0	3.174	0.4124	170.1
4	$T/3$	112.5^0	37.75^0	1.1336	40.0	3.5	0.3237	148.6
5	$5T/12$	139.95^0	44.75^0	0.6545	49.5	3.02	0.2166	127.64
6	$T/2$	150.0^0	47.75^0	0	50.0	3.0	0	100.00
7	$7T/12$	139.95^0	44.75^0	-0.6545	49.5	3.02	-0.2166	82.196285
8	$2T/3$	112.5^0	37.75^0	-1.1336	40.0	3.5	-0.3237	75.5458
9	$3T/4$	75.0^0	25.75^0	-1.3089	46.0	3.174	-0.4124	70.8014
10	$5T/6$	37.5^0	12.25^0	-1.1336	53.0	2.8869	-0.3925	71.8132
11	$11 T/12$	10.048^0	3.25^0	-0.6545	4.5	3.1978	-0.2046	83.0151
12	T	0^0	0^0	0	39.5	3.5316	0	100.00

Notes (a) The oscillating amplitude Ψ_{II} is 150^0, $T=1$ sec.

(b) $\psi_{II}^0 = \frac{1}{2}\left(1 - \cos\frac{2\pi}{T}t\right) \times 150^0$, unit degree

(c) $n_0 = \frac{\omega_f}{\omega_b} = \frac{\left(\pi \sin\frac{2\pi}{T}t\right) \times 150^0}{360^0}$, where $\omega_b = 360^0 / \sec$.

(d) values q_{bII}, ϕ_{II} ; see Fig. 5.3.

Draw line O_1O_2 first, then draw a series of lines from point O_2 making angles with line O_1O_2 according to successive ϕ_{II} in opposite direction, that is the reversal of the direction of the follower shaft rotation from the initial position.

On these lines, draw the point Q_j from the follower shaft center O_2

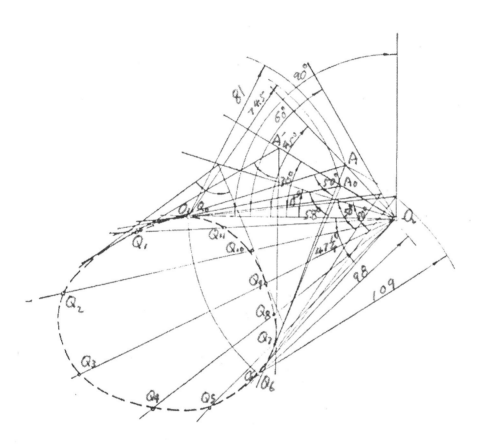

Figure 5.6 The centrode curve $Q\_\_Q_j$ for cam mechanism with favorable

transmission angle characteristics. Figures are not shown fully for clearness.

with the length of q_{bl} correspondingly one by one. Connect these Q_j's, then a

closed centrode curve $Q\_\_Q_j$ is completed, Fig.5.6 .

Centrode curve $Q\_\_Q_j$ is the key factor for the allowable minimum

transmission angle $[\mu_{min}]$ and the maximum and minimum radii of cam profile.

The angle between the follower O_2A and the frame O_1O_2 is another important factor. The larger is this angle O_1O_2A the smaller is the angle between the two lines both from point A tangent to centrode curve Q_Q_j; which must be so chosen where the two minimum transmission angles are nearly equal for the relationship $\Phi_{//} + \mu_{min1} + \mu_{min2} = 180^0$. The $\Phi_{//}$ is the very angle between the two tangent lines, since the follower of the first component mechanism is an integral part of the input link of the second component mechanism, whereas μ_{min1}, μ_{min2} are the minimum transmission angles on either side of the angle $\Phi_{//}$. On the other hand, the choices of the angle between the follower and the frame (i.e. $\angle O_1O_2A$) and the length of O_2 to A are not very effective. Fig.5.6 shows a series of O_1O_2A angles from $15^0 \rightarrow 90^0$. The angle subtended by the centrode curve Q_Q_j is changing as the angle O_1O_2A and the length of O_2A is being changed. We know from the figure that the subtended angle is about 50^0 when the angle $O_1O_2A = 90^0, 60^0$ or 45^0. Though the 50^0 is nice to have a larger sum of the angles of the two μ_{mn}'s, the point A is difficult to be found out for having two equal μ_{min}'s. Finally, we choose the angle $O_1O_2A = 45^0$, the subtended angle is also about 50^0. The μ_{min}'s are $63^0, 67^0$ each (not shown). The O_2A is 35 mm if O_1O_2 is 100 mm long. The choice of point A is by cut-and-try. It is not necessary to have two exactly equal μ_{min}'s for they are usually much larger than the μ_{mn}'s used for the second component mechanism. Though the point A has less influence on μ_{min}, the maximum and minimum radii of cam profile, r_{max}, r_{min}, are determined by the location of A when centrode curve Q_Q_j is determined.

The r_{max}, r_{min} are the distances from point A to the two intersecting points (i.e. O_1, O_1') which are the intersections of the arc, with radius O_1O_2 and center O_2, and the centrode curve Q_Q_j. A favorable cam we need should have a smaller r_{max} and the ratio between r_{max}, r_{min} not too large. The ratio r_{max}/r_{min} of this example is 109/81, about 1.34. If the angle $O_1O_2A' = 30^0$, the length of $O_2A' = 66$mm; a ratio of 108/54 (not figured), exactly 2, obtained is worse, though the two "μ_{min}"s is nearly equal to 60^0 each (not figured). The larger the

ratio is, the poorer the characteristics may be. Setting another point A_0, we get a

ratio 98/74.5=1.315. The ratio seems proper, but the difference of μ_{min} on both

sides is too big. It is quite a different job from the conventional cam mechanism

which the smallest radius r_{min} is considered for the motion of follower is regular

usually. Whereas the follower motion of the cam-link mechanism wholly depends

upon the output member of the second component mechanism, the cam profile of

the cam-link mechanism is hard to prescribe, so the r_{min} is not a decisive factor

should to be considered.

5.4 Cam Profile

Once the angle $O_1O_2A(= 45^0)$ and the length of O_2A (=35mm) are determined, the

cam profile can be found by conventional method. Unfortunately, the profile of

this example interferes with the follower in 4,5,6,7,8th positions. The successive

positions of O_2A shown in Fig.5.7 show that some positions

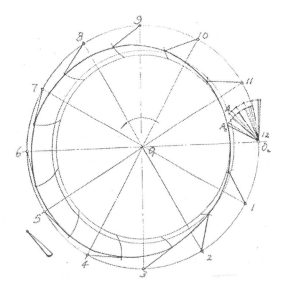

Figure 5.7   Cam profile design. Failed for $\angle O_1O_2A = 45^0, O_2A$=35 mm, profile

not shown. Modified for $\angle O_1O_2A = 30^0, O_2A$=30 mm.

exceed the circle of radius $O_1 O_2$. Changing angle $O_1 O_2 A$ to $O_1 O_2 A_0$ ($= 30^0$)and

reducing the length $O_2 A$ to $O_2 A_0$ (=30mm) may remedy the trouble (refer to

Fig.5.6). Furthermore, making the follower $O_2 A_0$ crooked instead of straight is

desired for keeping non-interference. After this deficiency is overcome , a new

problem arises. The $\mu_{\min}$ on the left side decreases to 40^0 which is less than 45^0,

the allowable transmission angle of the second component mechanism. We will
not consider this problem any more for it is not a decisive factor on this design.
For further study of cam profile, it is necessary to find out the radii and centers of
curvature in some positions needed.

(1) In Third Position of the Second Component Mechanism

$$\psi_{II3} = \frac{1}{2}\left(1 - \cos\frac{2\pi}{T} \times \frac{T}{4}\right) \times 150^0 = 75^0$$

$$\phi_{II3} = \psi_{I3} = 25.75^0$$

$$n_{03} = \frac{\omega_{f3}}{\omega_b} = \frac{\left(\pi \sin\dfrac{2\pi}{T} \times \dfrac{T}{4}\right) \times 150^0}{360^0} = 1.30899$$

$$q_{bII3} = 46\,\mathrm{mm}$$

$$q_{aII3} = a_{II} + q_{bII3} = 100 + 6 = 146$$

$$n_{II3} = \frac{q_{aII3}}{q_{bII3}} = \frac{146}{46} = 3.1739$$

$$n_{I3} = \frac{n_{03}}{n_{II3}} = 0.4124$$

By eq.(5.7), we have

$$q_{bI3} = \frac{a_I}{1 - n_{I3}} = 170.1 \quad \text{for } a_I = 100\,\mathrm{mm}$$

All the figures above, calculated or measured, can be found out in Table5.1.
Since the radii and centers of the curvature are calculated by the
acceleration of the mechanism output, so we may use the modified Hain's angular

acceleration equation for four-bar linkage (Appendix K) with graphical method to get the acceleration as follows:

$$\frac{\bar{\varepsilon}}{\omega^2} = \frac{\bar{a} \times \bar{m}}{q_b^{\,2}}$$

For the second component mechanism

$$\bar{A}_{II3} = \frac{\bar{\varepsilon}_{f3}}{\omega_{d3}^{\,2}} = \frac{\bar{a}_{II} \times \bar{m}_{f3}}{q_{bII3}^{\,2}} = \frac{100 \times 45}{46^2} = 2.1266\,(\text{C.C.W.})$$

where $m_{f3} = 45\,\text{mm}$ (measured from Fig.5.8.)

$$A_{03} = \frac{\varepsilon_{f3}}{\omega_b^{\,2}} = \frac{\left[2\pi^2 \cos\left(2\pi \times \frac{1}{4}\right)\right](\pi)\left(\frac{150^0}{180^0}\right)}{(2\pi)^2} = 0$$

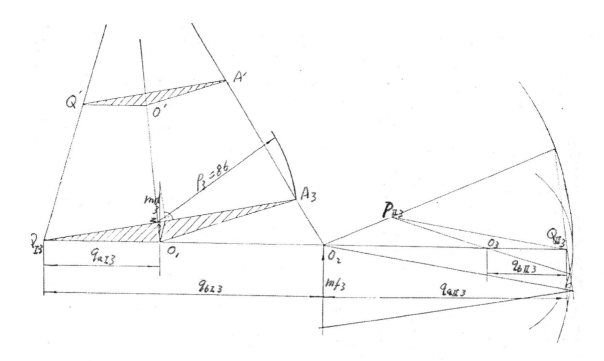

Figure 5.8   In third position

$$A_I n_{II} = -A_{II} n_I^{\,2}$$

$$\phi_{II3} = \psi_{I3} = 25.75^0$$

$$A_{I3} = \frac{-A_{II3}n_{I3}^2}{n_{II3}} = \frac{-(2.1266)(0.4124)^2}{3.174} = -0.1139 \ (\text{C.W.})$$

$$m_{d3} = \frac{-(0.1139)(170.1)^2}{100} = -32.9558 \ \text{mm}$$

Measuring the distance from O_2 to the left with the length of q_{bI3}

(=170.1mm) , we have Q_{I3}. Connect Q_{I3}, A_3, which presents the center line of the

coupler. From the end of the vector $\overline{m}_{d3}$ draw a line perpendicular to $Q_{I3}A_3$; and

from the foot draw a line to the head of vector $\overline{m}_{d3}$, this line should be parallel to

PQ (i.e. the collineation axis) , see Appendix K. Of course point P, the
instantaneous center of the coupler and frame, must be on the extension line of

O_2A_3. We should make an triangle $Q'A'O'$ ($\sim$ triangle $Q_{I3}A_3O_1$) instead of point

P's function as to the location of P is outside of this drawing. The triangle
$Q'A'O'$ can be located in a proper position on the drawing. Connecting points

O', O_1 ;intersecting line $Q_{I3}A_3$ at point Z____the center of curvature. We get the

radius of curvature ρ_3 (=86 mm).

(2) In Initial and Sixth Positions of the Second Component Mechanism

The intermediate member AO_2B is in its two limiting positions when the
second component mechanism is in its initial and sixth positions (Fig.5.9).
Simultaneously, the cam is in the dead center positions if the driving and driven
members are to be altered each other. The procedure for finding the centers and
lengths of ρ's in the two positions is as follows:

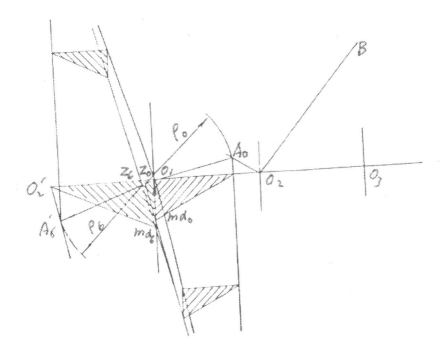

Figure 5.9   The initial and sixth positions of the second component mechanism, the dead center positions of the cam.

$$A_{00} = \frac{\varepsilon_{f0}}{\omega_b^{\,2}} = \left[\frac{2\pi^2 \cos\left(2\pi \times \dfrac{0}{T}\right)}{4\pi^2}\right]\pi\left(\frac{150^0}{180^0}\right) = 1.30899 \;(\text{C.C.W.})$$

$$A_{06} = \frac{\varepsilon_{f6}}{\omega_b^{\,2}} = \left[\frac{2\pi^2 \cos\left(2\pi \times \dfrac{T/2}{T}\right)}{4\pi^2}\right]\pi\left(\frac{150^0}{180^0}\right) = -1.30899 \;(\text{C.W.})$$

From Table 5.1, we have $n_{I0} = n_{I6} = 0$;

$$n_{II0} = 3.5316$$
$$n_{II6} = 3.0$$

Since $A_{00} = A_{I0}n_{II0} \; \therefore A_{I0} = \dfrac{1.30899}{3.5316} = 0.3706 \;(\text{C.C.W.})$

Also $A_{06} = A_{I6}n_{II6} \; \therefore A_{I6} = -\dfrac{1.30899}{3.0} = -0.4363\,(C.W.)$

Other parameters can be found in Table 5.1. Now we want to review some procedure for finding vector $\overline{m}$ and the "crankpin" location of the equivalent four-bar linkage.

When the four-bar linkage O_1ABO_2 is in its general position, Fig. 5.10, draw a line parallel to the colineation axis PQ from O_1 intersecting the coupler AB at S. From S draw a line perpendicular to AB intersecting the vertical line from O_1 (suppose O_1O_2 horizontal) at T. O_1T would be the vector $\overline{m}$ we needed. Furthermore, draw a line parallel to ST from Q. Then two similar triangles PQR and O_1ST present.

If the driving member O_1A (usually being a crank) coincides with the coupler AB, Fig.5.11, and the value of $\overline{m}$ is known by calculation, then the finding of crankpin A's location (i.e. the length of crank O_1A) becomes a problem as it is an equivalent four-bar linkage of a cam mechanism.

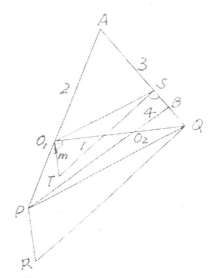

Figure 5.10 The finding of $\overline{m}$ of four-bar linkage O_1ABO_2 in a general position

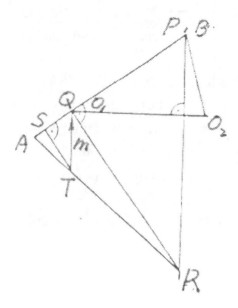

Figure 5.11   As an equivalent four-bar linkage of a cam mechanism, the layout
shows how the location of " crankpin" A is to be found.

Draw a line perpendicular to O_1O_2 from P (i.e. the intersecting point of
driving and driven members), and draw another line perpendicular to BQ (Q is
the intersecting point of coupler and frame) from Q. These two lines intersects at
R. From R draw a line passing through the tail (i.e. T) of vector $\overline{m}$ then
intersect the extension line of BQ at A. The reasons are: (1) points A,S,B,Q are
always in a straight line. Points A, O_1 , P are in another straight line. Sometimes
these two lines may be collinear; (2) $AS : AQ \equiv AO_1 : AP$. Thus the
"crankpin"____ the center of the curvature of the cam in its dead center position
presents. AB is the radius of the curvature certainly. This example is exactly the
same case of this description.

As A_I has been calculated, $q_{bl} = a_I = 100$ mm, then m_d in the initial and 6th

positions can be calculated, for

$$m_d = \frac{A_I \times q_{bl}^{2}}{a_I}$$

that is

$$m_{d0} = \frac{0.3706 \times 100^2}{100} = 37.06 \,\text{mm}$$

$$m_{d6} = \frac{-0.4363 \times 100^2}{100} = -43.63 \text{ mm}$$

Go back to Fig.5.9. Using the method above, from A_0 (equivalent to B in Fig.5.11) draw a line perpendicular to O_1O_2 intersecting the line which is perpendicular to O_1A_0 from O_1 at R. Passing this intersecting point R ,draw a line to the tail of vector $\overline{m}_{d0}$ intersecting the line A_0O_1's extension at Z_0. Then Z_0 is the center of curvature, Z_0A_0 is the radius of curvature ρ_0 which is equal to 85 mm.

It is evident that the follower moves about the cam in a direction opposite to the direction of rotation of the cam relatively. So the center of curvature Z_6 and the radius of curvature ρ_6 both can be obtained by the relative rotation method, see Fig.5.9.

It should be noted that both these two ρ's finding procedure would be outside of this drawing, so the two similar triangles are to be drawn as the lines perpendicular separately to O_1A_0, O_1A_6' can be drawn and the two Z's found.

(3) In Second Position of the Second Component Mechanism

This is a general position of the mechanism. Many parameters can be found in Table 5.1. Vector $\overline{m}_{f2}$ is obtained by drawing, Fig.5.12. It is just equal to zero by chance.

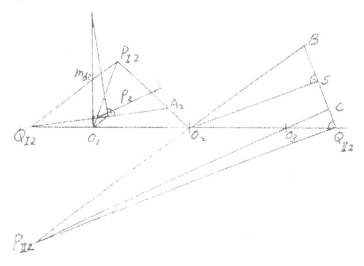

Figure 5.12   Second position, one of the general positions of the second
component mechanism.

Note: $m_{f2} = 0$

$$A_{II2} = \frac{\varepsilon_{f2}}{\omega_{d2}^2} - \frac{\varepsilon_{d2}}{\omega_b^2}\left(\frac{\omega_{f2}}{\omega_{d2}}\right) = \frac{a_{II} \times m_{f2}}{q_{bII2}} = \frac{100 \times 0}{53^2} = 0$$

$$A_{02} = \frac{\varepsilon_{f2}}{\omega_b^2} = \left[\frac{2\pi^2 \cos\left(2\pi\left(\frac{1}{6}\right)\right)}{(2\pi)^2}\right]\pi\left(\frac{150^0}{180^0}\right) = -1.1335 \qquad \text{(C.W.)}$$

$$A_{12} = \frac{\left(A_{02} - A_{II2}n_{12}^2\right)}{n_{II2}} = \frac{-1.1335 - (0)\left(0.3925^2\right)}{2.8869} = -0.3926 \quad \text{(C.W)}$$

$$A_{12} = \frac{a_I m_{d2}}{q_{bI2}} \; ;$$

$$-0.3926 = \frac{100 m_{d2}}{167.7^2} \; ;$$

$m_{d2} = - 110.4$ mm

$\rho_2 = 73$ mm

6 Came-Link Mechanism with Fixed Cam

6.1 Step Motion and Cam-Link Mechanism with Fixed Cam

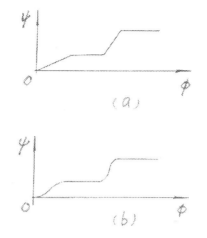

Figure 6.1   The step motion

The uniform velocity step motion is very popular in engineering circles.   Its input and output angular displacements are represented by ϕ, ψ. By observing it in Fig 6.1a , we see that there are violent shocks at the beginning and the end of every step. If a simple harmonic or a cycloidal projection motion is instead of the uniform velocity motion, the shocks at the beginning and at the end will greatly diminished.   The cam-link mechanism with fixed cam can be done fairly well for this purpose.

Take the mechanism in Fig.6.2 as an example. When crank O_1O_2 is rotating with a uniform velocity around O_2, link 3 (AO_1B) makes output link 5 do cycloidal projection motion due to the motion in fixed cam's groove by A .

No doubt it is a hard job to design the cam profile (i. e. The groove in which point A traces) and the dimension of the four-bar linkage O_1BCO_2. It would be easier to invert link 2 as a stationery frame, Fig.6.2b. The original frame (i. e. The fixed cam) becomes the driving member which   pushes point A. A typical three-member cam mechanism is to be composed. The four-bar linkage O_1BCO_2 has the same frame with the cam mechanism. Link O_1B connecting to follower AO_1 composes a cam-link mechanism in series. The inverted output link 5 (i. e. CO_2) makes relative oscillating motion to link 2. If they have the same angular speed; in other words, they have not any relative motion each other, the mechanism will dwell.

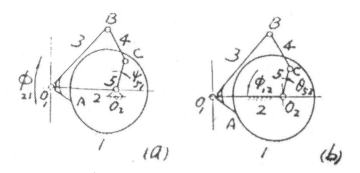

Figure 6.2   Cam-link mechanism with fixed cam: (a) configuration; (b) an
inversion of the mechanism, O_1O_2 being the frame

The pivot O_2 where all the links 1,2,5 rotate about is called compound
pivot. The instantaneous centers of the three links 12,25,51 are coincide at it. So
the sum of the three relative angular displacements around this pivot is zero, that
is

$$\phi_{21} + \psi_{15} + \theta_{52} = 0 \qquad\qquad (6.1)$$

where the subscripts are all relatives, for instance, ϕ_{21} is the angular displacement
of link 2 relative to link 1, the other relationships are the same as labeled.

The goal of this cam-link mechanism is the motion of link 5 relative to the

fixed cam 1, ψ_{51}. According to eq. (6.1) ,we have

$$-\psi_{15} = \phi_{21} + \theta_{52}$$

whereas $\psi_{51} = -\psi_{15}$, hence

$$\psi_{51} = \phi_{21} + \theta_{52} \qquad\qquad (6.2)$$

The sum of these relative angular velocities around this pivot is zero also,
that is

$$\omega_{21} + \omega_{15} + \omega_{52} = 0 \qquad\qquad (6.3)$$

Divide eq. (6.3) by ω_{21}

$$1 + \frac{\omega_{15}}{\omega_{21}} - \frac{\omega_{52}}{\omega_{21}} = 0$$

or

$$1 - \frac{\omega_{51}}{\omega_{21}} - \frac{\omega_{52}}{\omega_{12}} = 0$$

Let $\qquad n_0 = \frac{\omega_{51}}{\omega_{21}}, n_r = \frac{\omega_{52}}{\omega_{21}}$, then

$$n_0 = 1 - n_r \qquad\qquad\qquad (6.4)$$

This equation shows that when cam 1 is the base frame, the transmission ratio of the output link 5 to input link 2 will be determined by he transmission ratio of the output link 5 to "input cam " 1 if link 2 is inverted as the base frame.

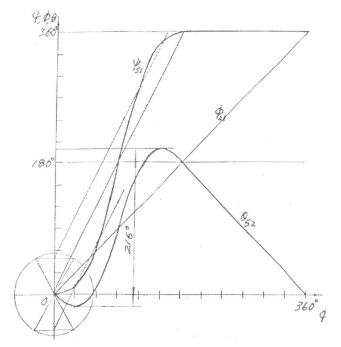

Figure 6.3   The relative motion characteristics among the chief members

In Fig.6.3 the relationships of motions show clearly. The motion of link 5 relative to link 2 , θ_{52} , is oscillation . The curve ψ_{51} is parallel to abscissa while link 5 dwells and the remains obey the cycloidal projection motion.

6.2 Linkage and Cam Profile Design

We have inverted the composite mechanism into a camed double rocker mechanism in above section.
Now we will discuss it in detail.

According to Appendix I when the ratio of output angle to the input angle of a double rocker mechanism is about 3, the allowable transmission angle $[\mu_{min}] = 30^0$ may be chosen. And the rest parameters are

$$L_1 : L_2 : L_3 : L_4 = 1 : 1.25 : 0.5 : \frac{\sqrt{3}}{4},$$ output angle $\psi \approx 217^0$, input angle $\phi = 67.5^0$.

As the required maximum oscillating angle $\theta_{52max} = 216^0$ just about 217^0 the provided value above. So all the parameters can be chosen directly, no more modifications is needed.

If one half input cycle of this step motion mechanism designed accomplishes the whole working stroke, then the following one half will dwell. Since the output motion in the working stroke wants to be the cycloidal projection motion, Fig. 6.3, the front half of the curve ψ_{51} is to be drawn accordingly. The rest is a straight line paralleling with abscissa. From eq. (6.2), we get

$$\theta_{52} = \psi_{51} - \phi_{21}$$

where the ψ_{51} has been drawn already, whereas ϕ_{21} is a inclined straight line making a 45^0 angle with the abscissa. Then curve θ_{52} is easy to draw by graphical method. The maximum amplitude is 216^0,

Curve θ_{52} is just the required motion of the member of inverted mechanism___ the oscillating follower after the inversion.

Let O_1O_2 is 100mm, then $O_1B = 125$ mm, BC=50mm, $CO_2 = 43.3$ mm. The oscillating amplitude period of link 5 (i.e. O_2C) is 180^0 of crank angle ϕ. Link 5 has a reverse motion up to 18^0 approximately about in first position, $\phi_{21} = 30^0$.

At the near end about in 5th position $\phi_{21} = 150^0, 18^0$ overtaking occurs.

Thus link 5 in first position is the lower limit, in 5th position is the upper limit. The upper limit of link 5 is the instant of that of link 3. The upper limit of link 5 may be obtained by cut-and-try. At about link 5 making an angle of 82^0 with horizontal, Fig. 6.4, the allowable minimum transmission angle $[\mu_{min}]$ is kept.

From this limit position draw $C_1, C_0, C_2, C_3, C_4, C_6, C_5$ on circumference C by the cycloidal projection motion (i.e. θ_{52}) respectively. In addition, some positions may be added for more accuracy. On circumference B , points $B_1, B_0, B_2, B_3, B_4, B_6, B_5$ can be drawn by the corresponding configurations of the mechanism.

The corresponding angular position of O_1B measured from horizontal are as follows:

$$(\phi_{31})_1 = 47^0, (\phi_{31})_0 = 41.5^0, (\phi_{31})_2 = 39^0, (\phi_{31})_3 = 13^0, (\phi_{31})_4 = -4^0, (\phi_{31})_6 = -7.5^0, (\phi_{31})_5 = -20^0$$

. As on circumference , there are $C_7, C_8, C_9, C_{10}, C_{11}$ equally spaced between $C_6 \rightarrow C_{12}$. And corresponding B, ϕ_{31} will be obtained.

The next step is to determine the length of AO_1 and the magnitude of angle AO_1B. These two are correlated. For instance the more the length of AO_1 the less the angle AO_1B could be. Since the lengths of O_1O_2, O_1B and the magnitude of angle $(\phi_{31})_1$ of the upper limit first position are defined,

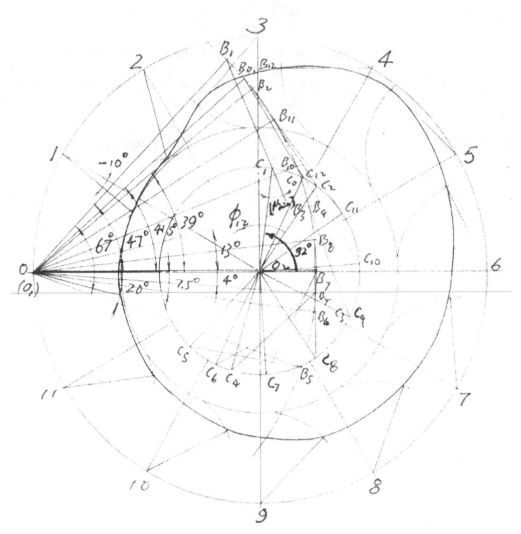

Figure 6.4   Design of linkage and cam profile

the only free choice parameter angle $\delta (\equiv \angle AO_1O_2)$ left for bewaring of jam in this position can be reduced as the length of AO_1 increases. In this example, we take $\angle \delta = -10^0$. So angle $AO_1B = (\phi_{31})_1 + |{-10^0}| = 47^0 + |{-10^0}| = 57^0$. Since the value of $(\phi_{31})_1$ is constant, however, the larger the length of AO_1 the less the angles δ, AO_1B will be if the jam characteristic is unchanged. The ratio between AO_1, O_1B is about 1 to 3 or 4. We take $AO_1 = 40\,\text{mm}$, just about one third of the length 125mm of O_1B. In first position, measuring the distance between O_2, A_1, we have the minimum radius of the cam, $r_{\min} = 61.0\,\text{mm}$. In fifth position, distance $O_2A_5 \equiv r_{\max}$ (=96.5mm).

We shall not follow the procedure as the above example. Just check the minimum transmission angle $\mu_{\min}$ in first position where the minimum radius is located, however, it is not exactly the minimum angle, just approximately. We find the angle interested by O_1A, O_2A is about 16^0 (not shown) in initial (0th) position. As this angle is too small (for it is the value close to, but not exactly, transmission angle) , a reformed angle $\delta(=-25^0)$ instead of -10^0 may be proper, so we can get about 40^0 (still not shown) instead of 16^0. Unfortunately, this modification presents another trouble. Both the follower's tip A and the cam profile would run-off the boundary of the circle forming by radius O_1O_2. It seems certain that a thorough redesign is necessary. A new double rocker mechanism for less $(\phi_{31})_1$ should be selected, however, this work would be left for readers. Now we continue to do the rest of design as if the trend of jam and low efficiency due to angle δ were not so severe for the oscillating follower. The cam profile shown in Fig.6.4 is a product of compromise. The extrema of θ_{52} represents analytically when $d\theta_{52}/d\phi_{21} = \omega_{52}/\omega_{21} = 0, \because \omega_{52} = 0; \therefore n_r = \dfrac{\omega_{52}}{\omega_{12}} = 0$. Then by eq. (6.4)

$$n_0 = 1 - n_r$$

we have

$$n_0 = 1$$

Parameter n_0 is the transmission ratio of link 5 to link 2. There are two tangential lines (not shown) on displacement curve ψ_{51} , Fig.6.3 ,being separately parallel to the slant line of ϕ_{21} where $n_0 = 1$.

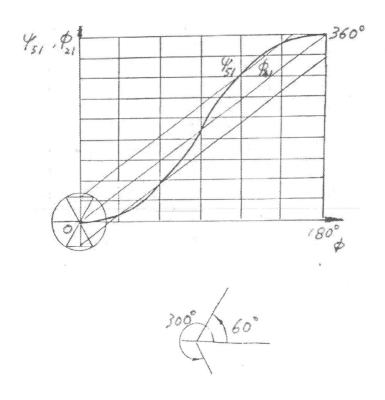

Figure 6.5   Cycloidal projection motion and $60^0, 300^0$

representations

The angular displacement of the cycloidal projection motion, Fig.6.5, is

$$\psi_{51} = 360^0 \left(\frac{\phi_{21}}{180^0} - \frac{1}{2\pi} \sin 2\pi \frac{\phi_{21}}{180^0} \right)$$ (6.5)

Differentiate it with respect to time and divided by ω_{21} , we have

$$\frac{\omega_{51}}{\omega_{21}} = n_0 = 360^0 \left[\frac{1}{180^0} - \frac{1}{2\pi} (\cos 2\pi \frac{\phi_{21}}{180^0}) \frac{2\pi}{180^0} \right]$$

for $n_0 = 1$ then

$$1 = 2 - 2 \cos 2\pi \left(\frac{\phi_{21}}{180^0} \right)$$

$$2 \cos 2\pi \left(\frac{\phi_{21}}{180^0} \right) = = 1$$

$$2 \phi_{21} = = 60^0, 300^0, \text{ see Fig.6.5,}$$

$$\phi_{21} = 30^0, 150^0$$

Substituting the values of ϕ_{21} into eq. (6.5) , we have

$$\psi_{51} = 360\left(\frac{30^0}{180^0} - \frac{1}{2\pi}\sin 2\pi(\frac{30^0}{180^0})\right) = 10..38^0$$

and $\qquad \psi_{51} = 360^0\left(\frac{150^0}{180^0} - \frac{1}{2\pi}\sin 2\pi(\frac{150^0}{180^0})\right) = 349.6181$

$$\theta_{52} = 10.38^0 - 30^0 = -19.62^0$$

and $\qquad \theta_{52} = 349.6181^0 - 150^0 = 199.6181^0$

then the oscillating angle is

$$199.6181^0 + \left|-19.62^0\right| = 219.23^0$$

Error percentage : $\quad \dfrac{219.23^0 - 216^0}{219.23^0} = 1.81\%$

7   Came-Link Mechanisms as Manipulators

7.1   Accessible   Regions for Two-Link Manipulators

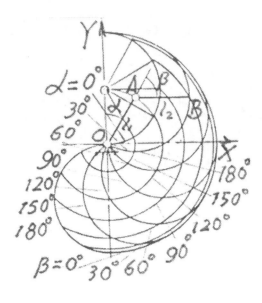

Figure 7.1   Accessible region for a two-link manipulator,

$$\frac{l_1}{l_2}=1$$

Assume the links 1,2 (connected in series) are the same (i.e. $l_1 = l_2$).   Restrict link 1's motion within the first and fourth quadrant, and the initial position coinciding with Y-axis. Angles α, β for links 1,2 are both measured clockwise (as positive) from Y-axis and link 1 respectively, ranging from $0^0 \rightarrow 180^0$. The tip B of link 2 is shown in the accessible region in Fig.7.1. Observing the area apart from the first and fourth quadrants is just equal to the empty area of first quadrant, therefore the total area of accessible region is exactly the half circle area with the diameter of $l_1 + l_2$. Furthermore, if we divide $0^0 \rightarrow 180^0$ of angle α into six equal parts, then the tip B revolving from $0^0 \rightarrow 180^0$ clockwise in each part presents an equal crescent accessible region.   Whereas angle β is different, the divided parts cover area less near $0^0, 180^0$; more near $\beta\left(= 90^0\right)$ though we divide from $0^0 \rightarrow 180^0$ of it is the same.

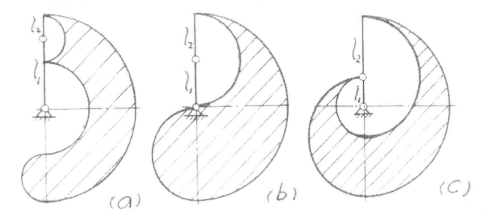

Figure 7.2　Comparison of the accessible regions of two-link

manipulator: $(a)\dfrac{l_2}{l_1}=\dfrac{1}{2};(b)\dfrac{l_2}{l_1}=1;(c)\dfrac{l_2}{l_1}=\dfrac{2}{1}$

Now we take the sum of lengths of links 1,and 2 as before, the ratios of link 2 to link 1 being 0.5,1, and 2 respectively. All the areas except Fig.7.2b are ring shapes, because the surplus can fulfill the blank as we mentioned above. The ring area can be calculated by the following equation, i.e.

$$A=\frac{\alpha}{2}\left(r_1{}^2-r_2{}^2\right)$$

$$(7.1)$$

where　$r_1=l_1+l_2, r_2=\left|l_1-l_2\right|$

The above equation is a special case, i.e. both angles α,β are equal to 180^0　The general equation should be as follows:

$$A=\frac{\alpha}{2}\left(r_1{}^2-r_2{}^2\right)$$

$$(7.2)$$

where　$r_1=l_1+l_2$

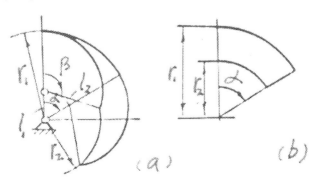

Figure 7.3 Two-link manipulator accessible region: $(a)\dfrac{l_2}{l_1} = 2.0; (b)$ equivalent

area

$$r_2 = \left(l_1^{\,2} + l_2^{\,2} + 2l_1 l_2 \cos\beta\right)^{\frac{1}{2}}$$

Let $\qquad A = \dfrac{\alpha r_1^{\,2}}{2}\left(1 - \dfrac{r_2^{\,2}}{r_1^{\,2}}\right)$

$$= \dfrac{\alpha(l_1 + l_2)^2}{2}\left[1 - \dfrac{1 + 2\dfrac{l_2}{l_1}\cos\beta + \dfrac{l_2^{\,2}}{l_1^{\,2}}}{\left(1 + \dfrac{l_2}{l_1}\right)^2}\right]$$

$$(7.3)$$

$$\dfrac{\partial A}{\partial \dfrac{l_2}{l_1}} = \dfrac{\left[1 + 2\dfrac{l_2}{l_1}\cos\beta + \left(\dfrac{l_2}{l_1}\right)^2\right] - \left(1 + \dfrac{l_2}{l_1}\right)\left(\cos\beta + \dfrac{l_2}{l_1}\right)}{\left(1 + \dfrac{l_2}{l_1}\right)^3}\left[\alpha(l_1 + l_2)^2\right]$$

$$(7.4)$$

Let $\dfrac{\partial A}{\partial \dfrac{l_2}{l_1}} = 0$, for finding the extrema. Then

$$1 + 2\dfrac{l_2}{l_1}\cos\beta + \left(\dfrac{l_2}{l_1}\right)^2 - \cos\beta - \dfrac{l_2}{l_1} - \dfrac{l_2}{l_1}\cos\beta - \left(\dfrac{l_2}{l_1}\right)^2 = 0$$

$$(1 - \cos\beta)\left(1 - \dfrac{l_2}{l_1}\right) = 0$$

When $\cos\beta = 1, \therefore \beta = 0 , A$ is extremum ($=0$)

When $l_2\big/l_1 = 1, A$ is extremum ($= A_{max}$).

Let $\qquad \dfrac{\partial A}{\partial \beta} = \dfrac{2\dfrac{l_2}{l_1}\sin\beta}{\left(1 + \dfrac{l_2}{l_1}\right)^2} = 0$, when $\beta = 0^0, 180^0 \therefore A = 0$.

Let $\quad \dfrac{\partial A}{\partial \alpha} = \dfrac{1}{2}(l_1 + l_2)^2 \left[1 - \dfrac{1 + 2\dfrac{l_2}{l_1}\cos\beta + \left(\dfrac{l_2}{l_1}\right)^2}{\left(1 + \dfrac{l_2}{l_1}\right)^2} \right] = 0$

then $\quad 1 + 2\dfrac{l_2}{l_1} + \left(\dfrac{l_2}{l_1}\right)^2 - 1 - 2\dfrac{l_2}{l_1}\cos\beta - \left(\dfrac{l_2}{l_1}\right)^2 = 0$

when $\quad \beta = 0^0 \therefore A = 0$.

Through the above discussion, following conclusions can be obtained.

（1）The extrema exist: when $\beta = 0^0, 180^0, \therefore A = 0$; when $\dfrac{l_2}{l_1} = 1, A = A_{max}$.

（2）A increases as β increases from $0^0 \rightarrow 90^0$; decreases as β increases from $90^0 \rightarrow 180^0$.

（3）The magnitude of A is determined by the absolute lengths of $l_1, l_2; \dfrac{l_2}{l_1}; (\alpha_{max} - \alpha_{min}); (\beta_{max} - \beta_{min})$; the location (by α); and the shape (by β).

[Example 7.1] If $l_1 = l_2 = 1, 0^0 \le \alpha \le 30^0, 0^0 \le \beta \le 180^0$. Compare the areas covered in six different intervals for 30^0 increments of β

$\vdots$

Solution

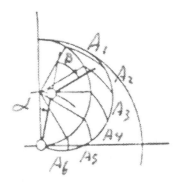

Figure 7.4 $\quad 0^0 \le \alpha \le 30^0$, divided crescents for 30^0 increments of β

$$A_1 = A_{01} = \frac{1}{2}\left(30^0\right)\left(\frac{\pi}{180^0}\right)\left(2^2 - 2 - 2\cos 30^0\right) = \frac{\pi}{12}(0.268) = 0.0701622$$

$$A_2 = A_{12} = \frac{\pi}{12}(1.0 - 0.268) = 0.1916371$$

$$A_3 = A_{23} = \frac{\pi}{12}(1.0) = 0.2617993$$

$$A_4 = A_{34} = \frac{\pi}{12}(1.0) = 0.2617993$$

$$A_5 = A_{45} = \frac{\pi}{12}(0.732) = 0.1916371$$

$$A_6 = A_{56} = \frac{\pi}{12}(0.268) = 0.0701622$$

We can use $A_3\left(= A_4\right)$ as an area unity for comparison. Then

$$A_1 : A_2 : A_3 : A_4 : A_5 : A_6 = 0.268 : 0.732 : 1.0 : 1.0 : 0.732 : 0.268$$

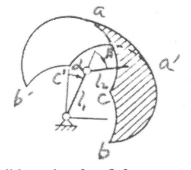

Figure 7.5   Accessible region for β from negative to positive (i.e.
$-120^0 \le \beta \le 120^0$) as an example, $l_2/l_1 = 1.0, 0 \le \alpha \le 60^0$

Besides, while angle β turns from $-120^0 \to +120^0$, Fig.7.5, the accessible

region a', b', c' (by $\beta = -120^0 \to 0^0$) is the mirror image of a, b ,c (by

$\beta = 0^0 \to 120^0$).

7.2 Synthesis on the Two-Link Manipulator

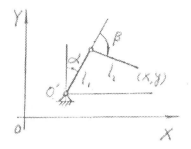

Figure 7.6   Two-link manipulator

In general, for easy programming the tip of two –link manipulator is restricted in the first quadrant, Fig.7.6.

[Example 7.2]   Design a two-link manipulator whose tip can touch the following working points:

$$K_1(90,20), K_2(80,15), K_3(100,30), K_4(110,40), K_5(120,20), K_6(130,10), K_7(100,20), K_8(100,10)$$

Solution:

Let the base point and the working area of the manipulator be in the first quadrant.

(1) Graphical method (Fig. 7.7)

(i)   Draw the prescribed array $K(x_K, y_K)$ K=1~8, one by one in the coordinate system.

(ii)   The base point $O_1(x_{01}, y_{01})$ may be arbitrarily chosen, so it can be the

origin point $O(x_0, y_0)$ of the coordinate.

(iii) Measure the maximum distance l_{max} from base point O_1 to the

farthermost point F ,the minimum distance l_{min} to the nearest point N Draw two

arcs, with base point O_1 as the center; l_{max}, l_{min} as the radii each; named

$\widehat{\beta}_{max}, \widehat{\beta}_{min}$.

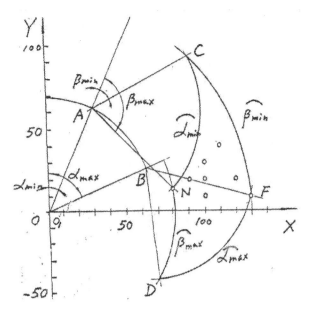

Figure 7.7   An example of two-link manipulator synthesis, graphical solution.

(iv) For avoidance of stretching of the two links in a straight line, the lengths l_1, l_2 should be equal to $l_{max}/1.9$ each. Draw an arc of radius l_1 , with O_1 as the center; then two arcs of radius l_2, with N and F as the centers separately, intersecting it with the arc at A and B. The line connecting the base point O_1 to A makes an angle with Y-axis, α_{min} ; and O_1B with Y-axis, α_{max} . Arc with AN (i.e. l_2) as base line and radius, A as center, intersecting arc $\hat{\beta}_{min}$ at C forms arc $\hat{\alpha}_{min}$.

The angles intersected by both AC and AN with OA's extension line are β_{min}, β_{max} respectively.

(v) Draw an arc with center B, radius l_2 intersecting arc $\hat{\beta}_{max}$ at point D ,arc $\hat{\beta}_{min}$ at point F, then arc $\hat{\alpha}_{max}$ is being presented. Now the accessible region bounded by these four arcs is presented.

(vi) Sometimes there are some points left outside of the accessible region, regulating the angles α or β's range may be adopted.

(vii) The accessible region can be calculated by planimeter or coordinate paper. For finding the optimum accessible region or link length, select some base point beside the origin of coordinates, compare the results, then the best one can be chosen.

The results of this example are:

$x_0 = 0, y_0 = 0, \alpha_{max} = 67^0, \alpha_{min} = 25^0, \beta_{max} = 107.5^0, \beta_{min} = 38^0, l_{max} = 131 \, \text{mm},$

$l_1 = l_2 = 68.9 \, \text{mm}, \quad A = 3768 \text{ sq. mm}$

(2) Analytical Method (Fig.7.8)

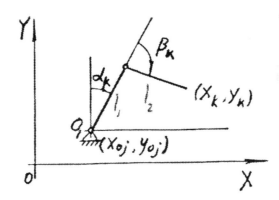

Figure 7.8 Two-link manipulator

Steps (i) and (ii) are the same as the graphical method. Step (iii) should use the following equation.

$$l_k = \sqrt{(x_k - x_{0j})^2 + (y_k - y_{0j})^2}$$

(7.5)

where k is the ordinal number of the array. Select l_{max}, l_{min}.

(iv) Calculate α_k, β_k by

$$\alpha_k = \cos^{-1} \frac{y_k - y_{0j}}{\sqrt{(x_k - x_{0j})^2 + (y_k - y_{oj})^2}} - \cos^{-1} \frac{(x_k - x_{0j})^2 + (y_k - y_{0j})^2 + l_1^2 - l_2^2}{2l_1 \sqrt{(x_k - x_{0j})^2 + (y_k - y_{0j})^2}} \qquad (7.6)$$

$$\beta_k = \cos^{-1} \frac{(x_k - x_{0j})^2 + (y_k - y_{0j})^2 - (l_1^2 + l_2^2)}{2l_1^2 l_2^2}$$

(7.7)

Select angle $\alpha_{max}, \alpha_{min}, \beta_{max}, \beta_{min}$. Then the accessible region can be

accomplished.

The equations may be simplified if the base point of the manipulator is the very point of coordinate origin. Then

$$\alpha_k = \cos^{-1}\frac{y_k}{\sqrt{{x_k}^2+{y_k}^2}} - \cos^{-1}\frac{\left({x_k}^2+{y_k}^2\right)+\left({l_1}^2-{l_2}^2\right)}{2l_1\sqrt{{x_k}^2+{y_k}^2}}$$

(7.8)

$$\beta_k = \cos^{-1}\frac{{x_k}^2+{y_k}^2-\left({l_1}^2+{l_2}^2\right)}{2l_1 l_2}$$

(7.9)

Furthermore, if $l_1 = l_2$ then

$$\alpha_k = \cos^{-1}\frac{y_k}{\sqrt{{x_k}^2+{y_k}^2}} - \cos^{-1}\frac{\sqrt{{x_k}^2+{y_k}^2}}{2l_1}$$

(7.10)

$$\beta_k = \cos^{-1}\frac{{x_k}^2+{y_k}^2-2{l_1}^2}{2{l_1}^2}$$

(7.11)

(v) Substitute the available parameters $\alpha_{max}, \alpha_{min}, \beta_{max}, \beta_{min}$, and the lengths of links into eq. (7.3) for computing accessible region A.

$$A = \frac{1}{2}\alpha(l_1+l_2)^2\left\{\left[1-\frac{1+2\frac{l_2}{l_1}\cos\beta_{max}+\left(\frac{l_2}{l_1}\right)^2}{\left(1+\frac{l_2}{l_1}\right)^2}\right]-\left[1-\frac{1+2\frac{l_2}{l_1}\cos\beta_{min}+\left(\frac{l_2}{l_1}\right)^2}{\left(1+\frac{l_2}{l_1}\right)^2}\right]\right\}$$

$$= \frac{1}{2}(\alpha_{max}-\alpha_{min})(l_1+l_2)^2\left\{\frac{2\frac{l_2}{l_1}[1-\cos\beta_{max}-(1-\cos\beta_{min})]}{\left(1+\frac{l_2}{l_1}\right)^2}\right\}$$

$$= (\alpha_{max}-\alpha_{min})(l_1+l_2)^2\left[\frac{\frac{l_2}{l_1}(\cos\beta_{min}-\cos\beta_{max})}{\left(1+\frac{l_2}{l_1}\right)^2}\right]$$

(7.12)

Assume the minimum of the accessible region A_{min} is needed, and the grid search technique is introduced. The results of this example can be obtained by computer. $A_{min} = 1897.551$ sq.mm

$L_{max} = 55.65986$ mm

$l_1 = l_2 = 29.2945$ mm

$x_{0j} = 77$ mm

$y_{0j} = 27$ mm

$\alpha_{max} = 89.588^0$

$\alpha_{min} = 15.754^0$

$\beta_{max} = 155.624^0$

$\beta_{min} = 36.389^0$

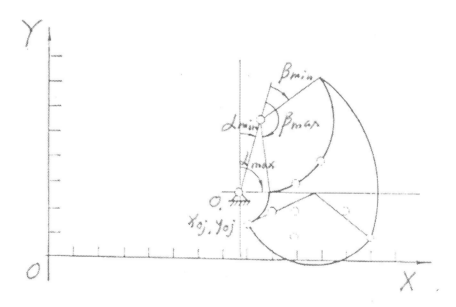

Figure 7.9 Two-link manipulator synthesis, analytical solution

Figure 7.9 shows the results. Something should be mentioned here: The grid net might be apart from the working points, neither too far nor too close, and should not enter the working zone definitely. The grid net area in this example is $x_{0j} = 0$ through 80 in X, $y_{0j} = 0$ through 80 in Y too; written as N $(0{\sim}80, 0{\sim}80)$. In the example we have studied the best base point (x_{0j}, y_{0j}) i.e. $O_1(77,27)$ is chosen. Other data will be shown for comparison, i.e.

(i) when grid net N ($0{\sim}70$) is used, the best base point will be O_1 (68,70), $A_{min} = 1927.565$ sq. mm

$l_{max} = 86.27862$ mm

(ii) when grid net $N(0\sim60,0\sim60)$ is used, the best base point will be $O_1(60,60)$,

$$A_{min} = 2129.479 \, \text{sq.mm}$$

$$l_{max} = 86.02325 \, \text{mm}$$

Comparing the above data, $N(0\sim80,0\sim80)$ is the best choice. $N(0\sim80,0\sim40)$ may be clever, however, for it can reduce the computing work to one half of the normal's.

7.3   Generation of Coupler Curves
_____ Application of Two-Link Manipulator Accessible Region Conception

The two-link manipulator in Fig. 7.10a can draw any curve in its accessible region, which is an open kinematic chain with two degrees of freedom. The cam-link mechanism in Fig.7.10b is a closed kinematic

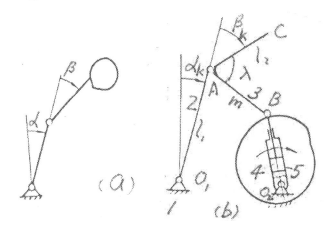

Figure 7.10 (a) Open kinematic chain; (b) the   cam-link mechanism which generates coupler curves

chain with one degree of freedom.

If $C(x_{ck}, y_{ck})$ are the sequence of points coordinate of the C's trace, then

that(i.e. the profile of the fixed cam) of the trace of $B(x_{bk}, y_{bk})$ is made according

to the demand of point C , with the help of links 2 and 3.

$$x_{bk} = l_1 \sin a_k + m \sin(a_k + \beta_k - \lambda) + x_{01}$$

$$(7.13)$$

$$y_{bk} = l_1 \cos a_k - m \cos(a_k - \beta_k - \lambda) - y_{01}$$

$$(7.14)$$

where $k = 1,2,3.......n$

n is the total number of the sequence of points on the trace of B.

$O_1(x_{01}, y_{01})$ is the base point of the maniplator which both coordinates might be

zeros. l_1, l_2 can be obtained by above section's method; α_k, β_k by eqs.

(7.6) ,(7.7); angle λ and length m arbitrarily chosen.

7.3.1 Synthesis on Stephenson Double- Cam Linkage

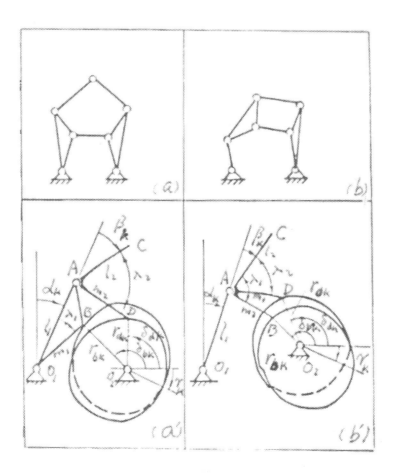

Figure 7.11   Stephenson double-cam linkage generates coupler curves. (a) and

(a'), original and its cam pair replacement (b) and (b') , inversion and its cam

pair replacement.

The original Stephenson mechanism and double-cam linkage are shown

inFig.7.11a. The follower tips B and D's traces are determined by point C's trace, a series of prescribed data $C(x_{ck}, y_{ck})$ The radii of cams, $r_b = BO_2, r_d = DO_2$, can be computed by the following equations.

$$r_{bk} = \sqrt{(x_{bk} - x_{02})^2 + (y_{bk} - y_{02})^2}$$

(7.15)

$$r_{dk} = \sqrt{(x_{dk} - x_{02})^2 + (y_{dk} - y_{02})^2}$$

(7.16)

where x_{02}, y_{02} are selected according to requirement.

$$x_{bk} = x_{01} + m_1 \sin(a_k - \lambda_1)$$

(7.17)

$$y_{bk} = y_{01} + m_1 \cos(a_k - \lambda_1)$$

(7.18)

$$x_{dk} = x_{01} - l_1 \sin a_k - m_2 \sin(a_k + \beta_k - \lambda_2)$$

(7.19)

$$y_{dk} = y_{01} + l_1 \cos a_k + m_2 \cos(a_k + \beta_k + \lambda_2)$$

(7.20)

$B(x_{bk}, y_{bk}), D(x_{dk}, y_{dk})$ are two series of data on the profiles of the two cams each.

Whereas the output cam angle positions are

$$\psi_{bk} = \delta_{bk} + \gamma_k = \tan^{-1} \frac{y_{bk} - y_{02}}{x_{bk} - x_{02}} + \gamma_k$$

(7.21)

$$\psi_{dk} = \delta_{dk} + \gamma_k = \tan^{-1} \frac{y_{dk} - y_{02}}{x_{dk} - x_{02}} + \gamma_k$$

(7.22)

where the δ_{bk}, δ_{dk} are the angles between the lines from tips B, D to cam shaft O_2 separately with the abscissa at the moment when C traces to k of certain value respectively; γ_k is the angular displacement from the initial position to this instant; $k=1,2.3......n$ n is the total number of the discrete points on the trace of C;

l_1, l_2, x_{01}, y_{01} can be obtained by the method in above section or arbitrarily chosen;

$m_1, m_2, \lambda_1, \lambda_2$ arbitrarily chosen too; α_k, β_k being computed by eqs.(7.6),(7.7).

The velocity of C can be controlled by the constant rotating sped of the cam shaft if no special demand is asked for.

7.3.2 Synthesis on Stephenson Double-Cam Linkage Inversion

The inversion and its cam pair replacement are shown in Fig.7.11b and b' respectively. Let $r_b = BO_2, r_d = DO_2$. The equations for finding $r_{bk}, r_{dk}, \psi_{bk}, \psi_{dk}$ are the same as above section, but

$$x_{bk} = x_{01} - l_1 \sin \alpha_k - m_1 \sin(\alpha_k - \beta_k - \lambda_1)$$

$$(7.23)$$

$$y_{bk} = y_{01} - l_1 \cos \alpha_k - m_1 \cos(\alpha_k - \beta_k - \lambda_1)$$

$$(7.24)$$

$$x_{dk} = x_{01} - l_1 \sin \alpha_k - m_2 \sin(\alpha_k - \beta_k - \lambda_2)$$

$$(7.25)$$

$$y_{dk} = y_{01} - l_1 \cos \alpha_k - m_2 \cos(\alpha_k - \beta_k - \lambda_2)$$

$$(7.26)$$

where l_1, l_2, x_{01}, y_{01} can be obtained by above method; $m_1, m_2, \lambda_1, \lambda_2$ being arbitrarily chosen; α_k, β_k being computed by eqs. (7.6) , (7.7).

7.3.3 Synthesis on Single-Cam Linkage

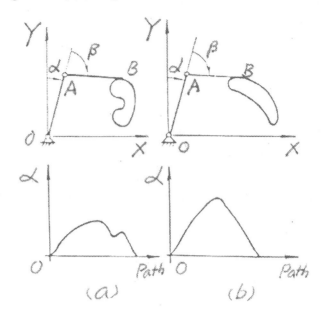

Figure 7.12   The oscillating of angle α: (a) oscillating twice; (b) oscillating once.

From Fig.7.12a we see that link OA cannot complete a prescribed motion with only one oscillation. On the contrary, in Fig.7.12b link OA oscillates just back and forth once, which can complete the prescribed motion. In the former case, a crank rocker mechanism must be taken as a controlling device. If a cam is fixed on the crank, then the motion of the link AB can be controlled by cam and crank simultaneously, Fig.7.13, for performing motion of C. This is an example of double-cam linkage replaced by single-cam linkage.

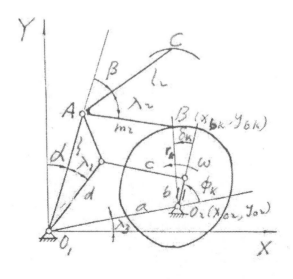

Figure 7.13   Single-cam linkage

The linkage used in the single-cam linkage is a crank rocker mechanism which can be designed according to optimum transmission angle, whereas its cam profile synthesis equations are

$$r_k = \sqrt{(x_{bk} - x_{02})^2 + (y_{bk} - y_{02})^2} \tag{7.27}$$

$$\delta_k = \tan^{-1} \frac{y_{bk} - y_{02}}{x_{bk} - x_{02}} - (\phi_k + \lambda_3) \tag{7.28}$$

where

r_k _____radius of cam

δ_k _____angular position relative to crank, positive for counterclockwise, negative

for clockwise

ϕ_k _____angular position of crank relative to frame

λ_3 _____angle between frame and abscissa

$$x_{bk} = x_{01} - l_1 \sin a_k - m_2 \sin(a_k - \beta_k - \lambda_2) \tag{7.29}$$

$$y_{bk} = y_{01} - l_1 \cos a_k - m_2 \cos(a_k - \beta_k - \lambda_2) \tag{7.30}$$

$$\lambda_3 = \tan^{-1} \frac{y_{02} - y_{01}}{x_{02} - x_{01}} \tag{7.31}$$

Now we would discuss the relationships among extrema of α (i.e. α_{ext}), α_k, ϕ_k.

Since there are two α_{ext}'s, when both links b, c lie on the same line, then one of

the two cases_____ while links b, c coincide will be

$$\alpha_{ext} = \frac{\pi}{2} - \lambda_1 - \lambda_3 - \cos^{-1} \frac{d^2 - a^2 - (c - b)^2}{2da} \tag{7.32}$$

For easy computing, we define $\alpha_k + \lambda_1 + \lambda_3 = \theta$, whereas

$$c^2 = [a\cos\lambda_3 + b\cos(\lambda_3 + \phi_k) - d\sin(\alpha_k + \lambda_1)]^2 + [a\sin\lambda_3 + b\sin(\lambda_3 + \phi_k) - d\cos(\alpha_k + \lambda_1)]^2$$

$$= a^2\cos^2\lambda_3 + b^2\cos^2(\lambda_3 + \phi_k) + 2ba\cos\lambda_3\cos(\lambda_3 + \phi_k) + d^2\sin^2(\alpha_k + \lambda_1)$$

$$- 2d\sin(\alpha_k + \lambda_1)[a\cos\lambda_3 + b^2\cos(\lambda_3 + \phi_k)] + a^2\sin^2\lambda_3 + b^2\sin(\lambda_3 + \phi_k)$$

$$+ 2ba\sin\lambda_3\sin(\lambda_3 + \phi_k) + d^2\cos^2(\alpha_k + \lambda_1) - 2d\cos(\alpha_k + \lambda_1)[a\sin\lambda_3 + b\sin(\lambda_3 + \phi_k)]$$

$$= a^2 + b^2 + d^2 + 2ba\cos\phi_k - 2da\sin\theta - 2db\sin(\theta + \phi_k);$$

and

$$\frac{a^2 + b^2 - c^2 + d^2}{2bd} + \frac{a}{d}\cos\phi_k - \frac{a}{b}\sin\theta - (\sin\theta\cos\phi_k + \cos\theta\sin\phi_k) = 0$$

Let

$$A = \frac{a^2 + b^2 - c^2 + d^2}{2bd} - \frac{a}{b}\sin\theta$$

$$B = \cos\theta$$

$$C = \frac{a}{d} - \sin\theta$$

then $\quad A - B\sin\phi_k + C\cos\phi_k = 0$

Let $\quad \tan\dfrac{\phi_k}{2} = t$, then

$$\sin\phi_k = \frac{2t}{1 + t^2}; \cos\phi_k = \frac{1 - t^2}{1 + t^2}$$

$$A - B\left(\frac{2t}{1 + t^2}\right) + C\left(\frac{1 - t^2}{1 + t^2}\right) = 0$$

$$(A - C)t^2 - 2Bt + (C + A) = 0$$

$$t = \frac{B \pm \sqrt{B^2 - (A^2 - C^2)}}{A - C}$$

$$. \quad \phi_K = 2\tan^{-1}\frac{B \pm \sqrt{B^2 - (A^2 - C^2)}}{A - C}$$

(7.33)

Since A, B, C are the functions of a, b, c, d, θ (i.e $\alpha_k, \lambda_1, \lambda_3$.) ; whereas all these parameters are prescribed except α_k, the relationship of $\phi = f(\alpha)$ is explicit.

Part III   Cam-Gear Mechanisms and the Others

8   Cam-Gear Mechanisms

8.1 Planetary Gear Train, Rack, and Cam Composition

Reviewing the planetary gear train, rack, and Scotch yoke composition, section 4.1, we know that the output pinion's cycloidal projection motion is satisfactory for both the accelerations at the beginning and the end of its motion cycle are equal to zeros. The deficiency of it is its inherent motion continuity (i.e. no intermittent motion is possible) formed by the simple harmonic motion of Scotch yoke and from the driving shaft by gear train. It is possible now at the designer's own will, to dwell or reverse, only if the Scotch yoke is replaced by cam. Fig. 8.1 is a typical cam-gear mechanism , whose degrees of freedom number

$$F = 3(7 - 1) - 2 \times 6 - 1 \times 5 = 1$$

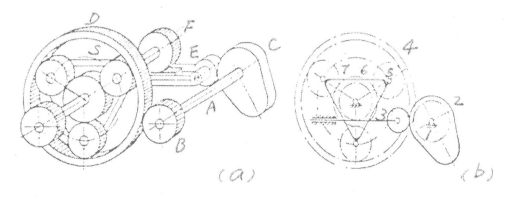

Figure 8.1   Planetary gear train, rack, and cam composition: (a) oblique drawing;(b) configuration

Being   attached to input shaft A and cam C , gear B is in mesh with the outer teeth D on ring gear 4 of the planetary gear train. The inner teeth of gear 4 is in mesh with the planet gear 5, and through gear 5 meshes with the central gear 7 simultaneously. The cam forces the rack E of follower 3 to rotate the meshing gear F. Gear F is integral with the cage 6. The planet gear 5 is pin-connected on the cage.

The uniformly rotating input motion is transmitted from gear B through the planetary gear train 4,5,6 to7. Besides, there is another source of motion which is from follower 3. This motion is very important, because the output motion of the pinion 7 is mainly controlled at the follower's command. The central pinion 7 may have certain characteristics; sometimes dwelling, and sometimes having zero accelerations at the beginning and the end of motion cycle. The following example will show the functions of this mechanism

.

[Example 8.1]

Let ϕ_1 = the angular displacement of cam shaft while the central pinion is at rest

ϕ_2 = the angular displacement of cam shaft while the central pinion is at work

$$K = \frac{2\pi}{\phi_2}$$

$$m_1 = \frac{N_B}{N_D}$$

$$m_2 = \frac{N_4}{N_7}$$

ω = constant angular velocity of camshaft

where N_B, N_D, N_4, N_7 are the teeth numbers of corresponding gears(N_4 is the inner teeth number of ring gear 4).

The cycloidal projection motion is to be the output central pinion motion because of the demands of zero accelerations at the beginning and the end of the cycle.

The kinematic characteristics of the central pinion

$$\psi = m_1 m_2 \left[K(\phi - \phi_1) - \sin K(\phi - \phi_1) \right] \tag{8.1}$$

$$\dot{\psi} = m_1 m_2 K\omega \left[1 - \cos K(\phi - \phi_1) \right]$$

$$\tag{8.2}$$

$$\ddot{\psi} = m_1 m_2 K^2 \omega^2 \sin K(\phi - \phi_1)$$

$$\tag{8.3}$$

where the ϕ is in the interval from $\phi_1 \to 2\pi$.

While the central pinion is in motion, the planetary gear train becomes a mechanism with two degrees of freedom, whose motion analysis is tabulated as follows:

Table 8.1

	Arm 6	Gear 4	Gear 7
Motion with arm relative to frame	ϕ_c	ϕ_c	ϕ_c
Motion relative to arm	0	X	$-m_2 X$
Total motion relative to frame	ϕ_c	$\phi_c + X$	$\phi_c - m_2 X$

According to the table above, we have

$$\phi_c + X = -m_1(\phi - \phi_1)$$

(1)

$$\phi_c - m_2 X = \psi = m_1 m_2 \left[K(\phi - \phi_1) - \sin K(\phi - \phi_1) \right]$$

(2)

$(1) \times m_2$ $\qquad \phi_c m_2 - X m_2 = - m_1 m_2 (\phi - \phi_1)$

(3)

$(2) - (3)$ $\qquad \phi_c(1 - m_2) = m_1 m_2 \left[K(\phi - \phi_1) - \sin K(\phi - \phi_1) \right] - m_1 m_2 (\phi - \phi_1)$ (4)

$$\phi_c = \frac{m_1 m_2}{m_2 + 1} \left[(\phi - \phi_1)(K - 1) - \sin K(\phi - \phi_1) \right]$$

(8.4)

This is the displacement equation of the cage while the output central pinion is at work. Differentiating it with respect to time and again for velocity and acceleration, we have

$$\dot{\phi}_c = \frac{m_1 m_2}{m_2 + 1} \omega \left\{ K \left[1 - \cos K(\phi - \phi_1) \right] - 1 \right\}$$

(8.5)

$$\ddot{\phi}_c = \frac{m_1 m_2}{m_2 + 1} \omega^2 K^2 \sin K(\phi - \phi_1)$$

(8.6)

where ϕ is in the interval from $\phi_1 \to 2\pi$.

While the central pinion is at rest, the planetary gear train is a one degree of freedom mechanism. The motion is tabulated as follows.

Table 8.2

	Arm 6	Gear 7	Gear 4
Motion with arm relative to frame	ϕ_c	ϕ_c	ϕ_c
Motion relative to arm	0	$-\phi_c$	ϕ_c / m_2
Total motion relative to frame	ϕ_c	0	$\phi_c + \dfrac{\phi_c}{m_2}$

From the above table, we have

$$\text{Displacement of arm 6 /displacement of gear 4} = \frac{\phi_c}{\phi_4} = \frac{\phi_c}{\phi_c + \dfrac{\phi_c}{m_2}} = \frac{m_2}{m_2 + 1}$$

$$\because \frac{\phi_4}{\phi} = -m_1$$

$$\therefore \phi_c = -\frac{m_1 m_2}{m_2 + 1} \phi$$

(8.7)

$$\dot{\phi}_c = -\frac{m_1 m_2}{m_2 + 1} \omega$$

(8.8)

$$\ddot{\phi}_c = 0$$

(8.9)

where ϕ is in the interval from $0^0 \rightarrow \phi_1$.

The kinematic characteristics of rack with follower

$$s = r\phi_c$$

(8.10)

$$v = r\dot{\phi}_c$$

(8.11)

$$a = r\ddot{\phi}_c$$

(8.12)

where r is the pitch radius of gear F.

The cam profile can be obtained according to s corresponding to the angular displacement of cam and the radius of base circle.

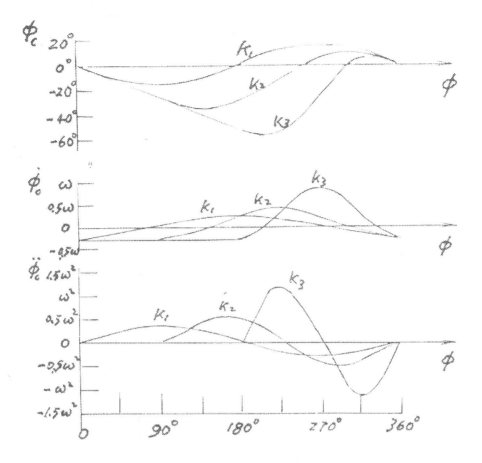

Figure 8.2   Kinematic characteristics of the cage of planetary gear train, rack ,
and cam composition;

$$K_1 = 1, K_2 = 1\frac{1}{3}, K_3 = 2$$

Figs.8.2 and 8.3 are the curves of angular displacements, angular velocities, and angular accelerations of the cage and output central pinion versus the input angular displacements of the cam shaft with different K values (i.e. $1,1\frac{1}{3},2$) or

$0^0, 90^0, 180^0$ dwell and $m_1m_2 = 1, m_2 = 2.5$

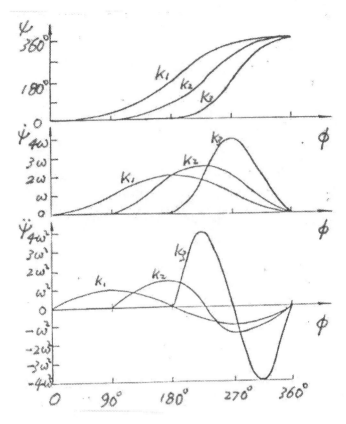

164

Figure 8.3   Kinematic characteristics of output central pinion of planetary gear train, rack, and cam composition.

$$K_1 = 1, K_2 = 1\frac{1}{3}, K_3 = 2$$

Besides, the same purpose can be accomplished if a oscillating gear sector follower is instead of the rack follower.

8.2 Worm Gear and Cylindrical Cam Composition

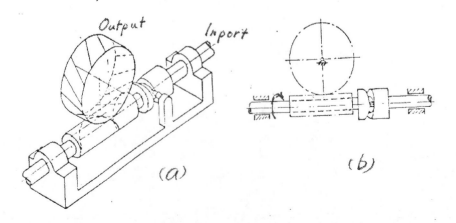

Figure 8.4   Cylindrical cam controlled worm and worm wheel mechanism: (a) oblique drawing; (b) configuration.

As shown in Fig.8.4 , the mechanism is an intermittent device controlled by a cylindrical cam. The cam is connected with worm as an integral. With the help of the roller on the fixed strut, the camshaft is being pushed along the shaft centerline. Then the rotation of input camshaft and its translation along centerline will control the output motion of the worm wheel in meshing with.

Generally speaking, the speed of input camshaft rotation is constant , whose translation is variable. The periodic motion of this composite mechanism is partly rotating and partly dwelling. For calculating convenience, parameters are to be listed conventionally as follows:

ϕ_1 = The angular displacement of input camshaft while the output wheel is at rest

ϕ_2 = The angular displacement of input camshaft while the output wheel is at work

$$K = \frac{2\pi}{\phi_2}$$

L = lead of worm

R = radius of pitch circle of output worm wheel

$$m = \frac{L}{2\pi R}$$

s_a = axial displacement of worm and cam

v_a = axial velocity of worm and cam

a_a = axial acceleration of worm and cam

s_{a1} = total axial displacement of worm and cam while the output wheel is at rest

Let the worm is right handed, rotating clockwise i.e. running over in front view. Besides, the output shaft has a conventional cycloidal projection motion. Then the kinematic characteristics of the output shaft

$$\psi = m[K(\phi - \phi_1) - \sin K(\phi - \phi_1)] \tag{8.13}$$
$$\dot{\psi} = mK\omega[1 - \cos K(\phi - \phi_1)] \tag{8.14}$$

$$\ddot{\psi} = mK^2\omega^2 \sin K(\phi - \phi_1) \tag{8.15}$$

where ϕ is in the interval from $\phi_1 \rightarrow 2\pi$

The axial translation of worm and cam while the output wheel is at rest,

$$s_a = -\frac{L\phi}{2\pi}, \left(s_{a1} = -\frac{L\phi_1}{2\pi} \because \phi = \phi_1\right) \qquad (8.16)$$

$$v_a = -\frac{L\omega}{2\pi} \qquad (8.17)$$

$$a_a = 0 \qquad (8.18)$$

where ϕ is in the interval from $0^0 \to \phi_1$.

The axial translation of worm and cam while the output wheel is at work

$$s_a = mR\left[K(\phi - \phi_1) - \sin K(\phi - \phi_1)\right] - \frac{L}{2\pi}(\phi - \phi_1) \qquad (8.19)$$

$$v_a = mRK\omega\left[1 - \cos K(\phi - \phi_1)\right] - \frac{L\alpha}{2\pi} \qquad (8.20)$$

$$a_a = mRK^2\omega^2 \sin K(\phi - \phi_1) \qquad (8.21)$$

where ϕ is in the interval from $\phi_1 \to 2\pi$.

The data used for design of cylindrical cam is provided by the axial displacement s_a of worm and cam.

9   The Planet Gear and Geneva Wheel Composition

9.1 Introduction

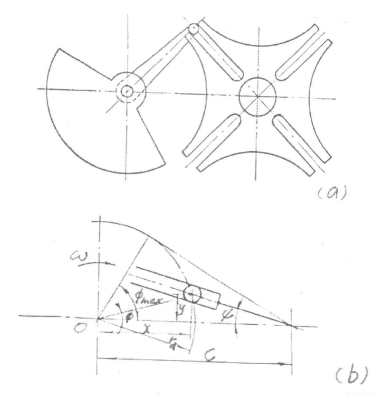

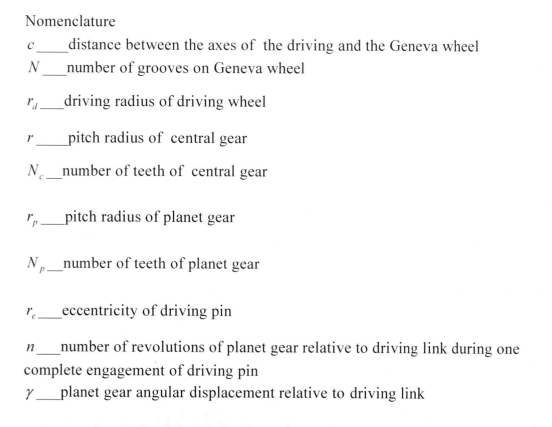

Figure 9.1   Standard Geneva   wheel: (a) construction drawing; (b) geometric relationships

Nomenclature

c ____distance between the axes of the driving and the Geneva wheel

N ____number of grooves on Geneva wheel

r_d ____driving radius of driving wheel

r ____pitch radius of central gear

N_c ____number of teeth of central gear

r_p ____pitch radius of planet gear

N_p ____number of teeth of planet gear

r_e ____eccentricity of driving pin

n ____number of revolutions of planet gear relative to driving link during one complete engagement of driving pin

γ ____planet gear angular displacement relative to driving link

ϕ ___driving shaft angle

ϕ_{max} _angle ϕ at start of engagement

ψ ___angular displacement of Geneva wheel

$\dot{\psi}$ ___angular velocity of Geneva wheel

$\ddot{\psi}$ __ angular acceleration of Geneva wheel

ω ___angular velocity of driving link

The Geneva wheel generally used for indexing has a shockless performance; that is the moment when the driving pin is entering the groove, its tangential direction coincides with the groove centerline. Though it is shockless, whose acceleration is not zero, it still has infinite jerk exists. In other words, the rate of change of acceleration is infinity. When the driving wheel rotates with a uniform speed, the driving pin inherently has a centripetal acceleration $r_d \omega^2$, Fig.9.1. At the first moment, the angular acceleration component applied to Geneva wheel groove by the driving pin is $\omega^2 ctn\phi_{max}$,i.e. $\ddot{\psi}$. As

$$\psi = \tan^{-1} \frac{y}{c-x} = \tan^{-1} \frac{r_d \sin\phi}{c - r_d \cos\phi} \tag{9.1}$$

Differentiating it with respect to time for angular velocity and again for angular acceleration, we have

$$\dot{\psi} = \frac{\left(cr_d \cos\phi - r_d^2\right)\omega}{c^2 - 2cr_d \cos\phi + r_d^2}$$

$$\tag{9.2}$$

$$\ddot{\psi} = \frac{cr_d \sin\phi\left(r_d^2 - c^2\right)\omega^2}{\left(c^2 - 2cr_d \cos\phi + r_d^2\right)^2}$$

$$\tag{9.3}$$

The angular acceleration $\ddot{\psi}$ might cause violent jerk and destroy he Geneva wheel. Since ϕ_{max} decreases as the number of grooves decreases, the angular acceleration of the Geneva wheel has to be raised. That is: ϕ_{max} decreases, $ctn\phi_{max}$ will increase, and $\ddot{\psi}$ will increase too for $\ddot{\psi} = \omega^2 ctn\phi_{max}$ .For the avoidance of jerk, the angular acceleration must be zero at the beginning and the end while the driving pin engages the Geneva wheel in moving. This can be done by planet-Geneva mechanism.

Fig.9.2 shows he planet-Geneva mechanism. The central gear and the driving link are coaxial. The planet shaft and the central gear axis are connected by the driving link. Since the driving pin is fixed on the planet gear and located off the planet gear pitch circle, which generates trochoid. The design demand is that: at the beginning and the end of the engagement the driving pin has no acceleration component in the direction perpendicular to the groove centerline. That is to say, at the beginning and the end of engagement the center of the driving pin must be at the center line of driving link and between the central gear shaft center and planet gear shaft center.

Now we are going to compute the angular acceleration $\ddot{\psi}$ of this mechanism. Let γ be the angular displacement of the planet gear relative to the driving link. The ratio of the angular displacement of the driving link to that of the planet gear relative to the driving link is

$$\frac{\phi_{max}-\phi}{\gamma}=\frac{1}{\dfrac{r}{r_p}}$$

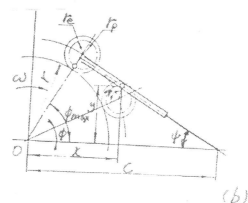

(a)

(b)

Figure 9.2 Planet-Geneva mechanism: (a) configuration; (b) geometric relationships

After the revolving of ($\phi_{max} - \phi$) of the driving link, calculating according to the planetary gear train principle , we have

$$(\gamma - \phi) = \frac{r}{r_p}\phi_{max} - \left(\frac{r}{r_p} + 1\right)\phi$$

$$\psi = \tan^{-1}\frac{y}{c-x} = \tan^{-1}\frac{(r+r_p)\sin\phi + r_e\,\sin(\gamma-\phi)}{c-(r+r_p)\cos\phi + r_d\,\cos(\gamma-\phi)}$$

(9.4)

$$\frac{\dot{\psi}}{\omega} = \frac{x\dfrac{dy}{d\phi} - y\dfrac{dx}{d\phi}}{x^2 + y^2}$$

(9.5)

$$\frac{\ddot{\psi}}{\omega^2} = \frac{x\dfrac{d^2y}{d\phi^2} - y\dfrac{d^2x}{d\phi^2}}{x^2 + y^2} - \frac{2xy\left[\left(\dfrac{dy}{d\phi}\right)^2 - \left(\dfrac{dx}{d\phi}\right)^2\right] + 2(x^2 - y^2)\left(\dfrac{dx}{d\phi}\right)\left(\dfrac{dy}{d\phi}\right)}{\left(x^2 + y^2\right)^2}$$

(9.6)

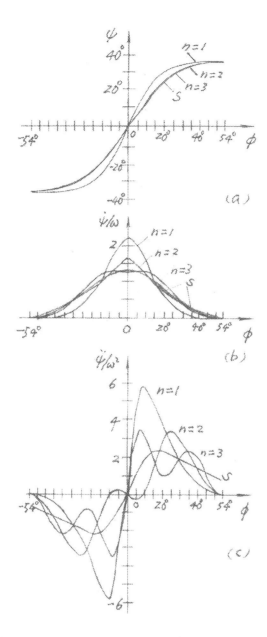

Figure 9.3   Kinematic characteristics of the five-groove standard Geneva wheel mechanisms and that of planet-Geneva mechanism with planet gears, n =1,2,3.

Curves of(a) angular displacement ψ ;(b) specific angular velocity $\dot{\psi}/\omega$ and(c) specific angular acceleration $\ddot{\psi}/\omega^2$ versus driving link input angle ϕ

 For number of grooves N=3,4,5; revolving number of planet gear relative to the driving link n=1,2,3; every 2.5^0 increment of input angle ϕ; the angular displacements, velocities , accelerations of the standard Geneva wheel and the planetary-Geneva mechanism are tabulated in Referance [1]. Its typical characteristics is shown in Fig.9.3.

9.2 Design

For the angular acceleration of the Geneva wheel of this mechanism has to be zero at the beginning and the end of engagement, the origin of the moving translation coordinate must be located on the planet gear shaft. Then the absolute acceleration of the driving pin

$$\bar{a}_a = \bar{a}_e + \bar{a}_r$$

where $\bar{a}_e$ is the acceleration of the translation coordinate, $\bar{a}_r$ is the acceleration relative to the translation coordinate. Since the driving link rotates with a uniform speed and the translation coordinate has no angular speed, therefore

$$\bar{a}_e = \bar{a}_{en}, a_{en} = (r + r_p)\omega^2$$

$$\bar{a}_r = a_{rn}, a_{rn} = r_e\left[\left(\frac{r}{r_p} + 1\right)\omega\right]^2$$

where $\left(\dfrac{r}{r_p} + 1\right)\omega$ is the angular velocity of the driving pin relative to the translation coordinate.

If we want $\bar{a}_a = 0$, let $\bar{a}_{en} = -\bar{a}_{rn}$, then

$$\bar{a}_{en} + \bar{a}_{rn} = 0$$

$$-(r + r_p)\omega^2 + r_e\left[\left(\frac{r}{r_p} + 1\right)\omega\right]^2 = 0$$

$$r_e = \frac{r_p^2}{r + r_p}$$

$$(9.7)$$

Since the driving pin at the beginning and the end of the engagement has to be on the centerline of the driving link and between centers of central gear and planet, the arc length of planet gear rolling one revolution on the stationary central gear (i.e. the circumference of the planet gear) is just equal to the arc length between two adjacent grooves, i.e.

$$2\pi r_p n = r(2\phi_{max}) \text{ ,then}$$

$$2\phi_{max} = \pi - \frac{2\pi}{N}$$

$$\therefore r_p = \frac{N-2}{2nN}r$$

(9.8)

Substituting it into eq.(9.7), yield

$$r_e = \frac{(N-2)^2 r}{2nN(2nN+N-2)}$$

(9.9)

and $\qquad \phi_{max} = \frac{N-2}{2N}\pi$

(9.10)

whereas

$$c = \frac{1}{\cos\phi_{max}}\left[1 + \frac{N-2}{2nN} - \frac{(N-2)^2}{2nN(2nN+N-2)}\right]r$$

(9.11)

[Example 9.1] Design an index device, $N=4, n=3, c \approx 15\,cm$

Solution:

$$\phi_{max} = \frac{N-2}{2N}\pi = \frac{\pi}{4} = 45^0$$

$$c = \frac{1}{\cos\phi_{max}}\left[1 + \frac{N-2}{2nN} - \frac{(N-2)^2}{2nN(2nN+N-2)}\right]r = 1.53r$$

$$r = \frac{15}{1.53} = 9.79$$

Let $r = 10\,cm$

$r_p = \frac{N-2}{2nN}r = 0.83\,cm$. Let $r_p = 0.8\,cm$.

And again let $m = 0.8\,mm$, the number of teeth on planet gear

$$N_p = \frac{r_p \times 2}{m} = 20$$

$$r_e = \frac{(N-2)^2 r_p}{2nN[2nN+N-2]} = 0.041\,cm$$

$\because r = \frac{mN_c}{2}$, let $m = 0.8\,mm$, then the number of teeth on central gear

$$N_c = \frac{2r}{m} = 250$$

c =15.3 cm

9.3 Discussion

Though the kinematic characteristics shown in Fig.9.3 merely belongs to a five-groove standard Geneva wheel and planet Geneva mechanism, which is worth studying for being the reference of the three , four, and six-groove Geneva wheels.

First ,observe that the angular displacements of all the planet-Geneva mechanisms, except $n=1$, are similar to simple Geneva wheel. The three curves are so similar that they may seem as if a single one.

Secondly, observe the ratio of specific angular velocity, i.e. $\dot{\psi}/\omega$, versus the input crank angle ϕ, which is as high as 2.2087 as for $n=1$, whereas 1.42592 as for s——the five-groove standard Geneva wheel. The difference is great. However, the difference in accelerations would be much more. The maximum $\ddot{\psi}/\omega^2$ is 5.7324 as for $n=1$, 2.29880 for s .

It must be noted that when $n=2$, the angular acceleration of the planet-Geneva mechanism has frequent sense changing from positive to negative twice and negative to positive once in a complete motion cycle. So there are three shocks in one cycle. However, both at the beginning and the end of the motion cycle no shock occurs (i.e. $\ddot{\psi} = 0$). Whereas $n=1$ or 3, only one shock occurs. Therefore, $n=2$ should be avoid if 5 or 6 grooves of planet-Geneva mechanism is used. But $n=2$ can be used if 3 or 4 grooves is used. These are all determined by the characteristics of the mechanisms.

10   Geneva Wheels Connected in Series

10.1 Introduction

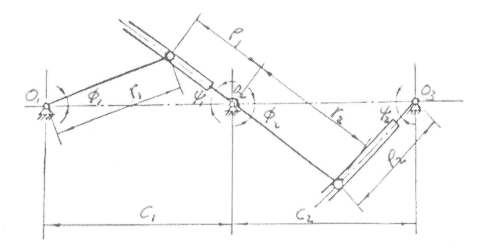

Figure 10.1   Geneva wheels connected in series; left to right , primary to

  secondary

The Geneva wheels connected in series, Fig.10.1 , might avoid an infinite jerk, only

  a finite amount remains on the output shaft, for the velocity and even

  acceleration on the wheel are both zeros when the driving pin enters and leaves

  the wheel groove. The primary input crank pushes the primary Geneva wheel,

  the Geneva wheel is integral with the secondary crank. So the secondary crank

  drives the secondary Geneva wheel completing the output motion of this

  composite mechanism. If we set the output shaft coaxial with the input shaft,

  the mechanism is an inverted one, the total geometric shafts reduce to two.

10.2 Output Angular Velocity

  Let the output angular velocity of this composite mechanism is

$$\omega_3 = \frac{d\psi_2}{dt} = \frac{d\psi_2}{d\phi_2}\frac{d\phi_2}{dt}, \because \frac{d\phi_2}{dt} = \frac{d\psi_1}{dt},$$

$$\therefore \omega_3 = \frac{d\psi_2}{d\phi_2}\frac{d\psi_1}{d\phi_1}\frac{d\phi_1}{dt}$$

If the input angular velocity is ω_1, which is equal to $\dfrac{d\phi_1}{dt}$ then

$$\omega_3 = n_I n_{II} \omega$$

$$(10.1)$$

where n_I, n_{II} are transmission ratios of the primary and secondary component

mechanisms respectively, and

$$n_I = \frac{d\psi_1}{d\phi_1} = \frac{d}{d\phi_1} \sin^{-1}\left(\frac{r_1}{\rho_1} \sin \phi_1 \right) = \frac{c_1 r_1 \left(\cos \phi_1 - \dfrac{r_1}{c_1} \right)}{\rho_1^2} \qquad (10.2)$$

$$n_{II} = \frac{d\psi_2}{d\phi_2} = \frac{d}{d\phi_2} \sin^{-1}\left(\frac{r_2}{\rho_2} \sin \phi_2 \right) = \frac{c_2 r_2 \left(\cos \phi_2 - \dfrac{r_2}{c_2} \right)}{\rho^2_2} \qquad (10.3)$$

where $\rho = \sqrt{c^2 + r^2 - 2cr \cos \phi}$

10.3   Output Angular Acceleration

The output angular acceleration of this mechanism

$$\varepsilon_3 = \frac{d\omega_3}{dt} = d(n_I n_{II} \omega_1)$$

The input angular velocity is constant conventionally, then

$$\varepsilon_3 = \omega_1\left(n_I \frac{dn_{II}}{dt} + n_{II} \frac{dn_I}{dt} \right)$$

$$= \omega_1\left(n_I \frac{dn_{II}}{d\phi_2} \frac{d\psi_1}{d\phi_1} \frac{d\phi_1}{dt} + n_{II} \frac{dn_I}{d\phi_1} \frac{d\phi_1}{dt} \right)$$

$$= \omega_1^2\left(n_I^2 A_{II} + n_{II} A_I \right) \qquad (10.4)$$

where

$$A_I = \frac{dn_I}{d\phi_1} = \frac{c_1^4\left[\left(\dfrac{r_1}{c_1} \right)^3 - \dfrac{r_1}{c_1} \right] \sin \phi_1}{\rho_1^4} = \frac{1}{\rho_1^4}\left(c_1 r_1^3 - c_1^3 r_1 \right)\sin \phi_1 \qquad (10.5)$$

$$A_{II} = \frac{dn_{II}}{d\phi_2} = \frac{c_2^{\ 4}\left[\left(\dfrac{r_2}{c_2}\right)^3 - \dfrac{r_2}{c_2}\right]\sin\phi_2}{\rho_2^{\ 4}} = \frac{1}{\rho_2^{\ 4}}\left(c_2 r_2^{\ 3} - c_2^{\ 3} r_2\right)\sin\phi_2 \tag{10.6}$$

10.4 Conditions of Zero Angular Accelerations at the Beginning and the End of Intermittent Output Motion

The angular velocity of primary Geneva wheel is

$$\omega_2 = \frac{d\psi_1}{dt} = \frac{d\psi_1}{d\phi_1}\frac{d\phi_1}{dt} = n_I \omega_1 = \frac{c_1 r_1\left(\cos\phi_1 - \dfrac{r_1}{c_1}\right)}{\rho_1^{\ 2}}\omega_1 \tag{10.7}$$

The angular acceleration of primary Geneva wheel is

$$\varepsilon_2 = \omega_1^{\ 2}\left(\frac{c_1 r_1^{\ 3} - c_1^{\ 3} r_1}{\rho_1^{\ 4}}\right)\sin\phi_1 \tag{10.8}$$

Fig.10.2 shows a single Geneva wheel's kinematic characteristics. The specific angular velocities are equal to zeros both at the beginning and the end of the engagement, whereas the specific angular accelerations are not zeros, namely

$$\frac{d\psi_2}{d\phi_1} = \frac{\omega_3}{\omega_1} = 0,$$

$$\frac{d^2\psi_2}{d\phi_1^{\ 2}} = \frac{\varepsilon_2}{\omega^2} \neq 0$$

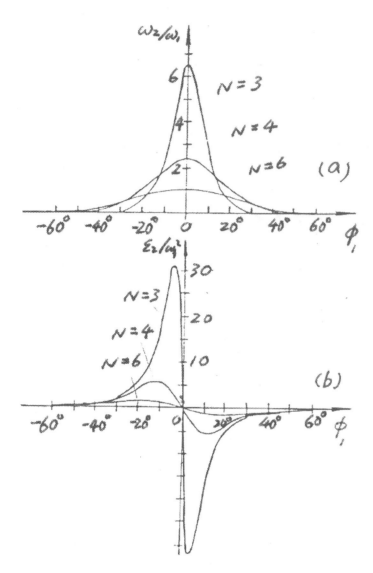

Figure 10.2   Kinematic characteristics of Geneva wheel, $N = 3, 4, 6$: (a) the

specific angular velocity $\omega_2 \big/ \omega_1$ versus ϕ_1; (b) the specific angular acceleration

$\varepsilon_2 \big/ \omega_1^{\,2}$ versus ϕ_1

If two Geneva wheels are connected in series, all the angular velocities and angular

accelerations are equal to zeros both at the beginning and the end of the crank pin

engagement in certain conditions. Look at eq.(10.4), as $\varepsilon_3 = \omega_1^{\,2}\left(n_I^{\,2} A_{II} + n_{II} A_I \right)$;

assume the angular velocities of primary Geneva wheel are zeros both at the

beginning and the end of engagement, n_I will be zero and so does the secondary

Geneva wheel $n_{II} = 0$ too. Then ε_3 will be zero. There are three combinations which can meet this requirement. Fig.10.3 and Table 10.1 show these combinations and their maxima of angular velocities and accelerations.

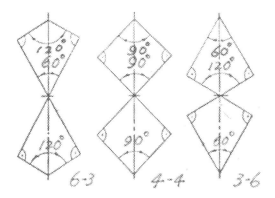

Figure 10.3   Three combinations of primary and secondary Geneva wheels which

can have $\varepsilon_3 = 0$

Table 10.1   The Maximum Specific Angular Velocities and Specific Angular

Accelerations of the Output Shaft of the Three Combinations

No. of grooves of primary Geneva wheel	No. of grooves of secondary Geneva wheel	$\dfrac{\omega_{3(max)}}{\omega_1}$	$\dfrac{\varepsilon_{3(max)}}{\omega_1^2}$
6	3	$1 \times 6.46 = 6.46$	32.42
4	4	$2.41 \times 2.41 = 5.85$	33.88
3	6	$6.46 \times 1 = 6.46$	62.25

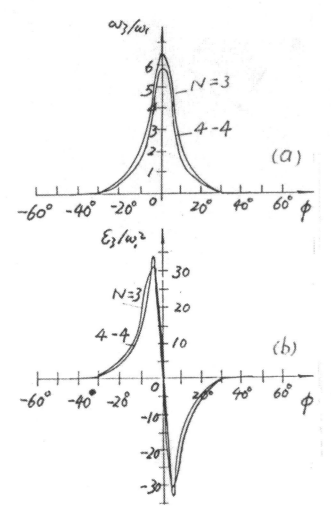

Figure 10.4　Kinematic characteristics comparisons between the single three-groove Geneva wheel and 4-4 combination: (a). ω_3/ω_1 versus ϕ_1 ;(b) ε_3/ω_1^2 versus ϕ_1

Fig. 10.4 show the curves of ω_3/ω_1, ε_3/ω_1^2 for the three-groove Geneva wheel and

4-4 combinations. Since the number of grooves combination of 3-6 has is

specific angular acceleration, i.e. ε_3/ω_1^2 , as high as 62.25, it should be rejected,

however. Though the characteristics of angular velocities and accelerations of

the 6-3 and 4-4 combinations are very close to that of single three -groove

Geneva wheel, their angular accelerations at the beginning and the end of engagement are vanished, a smooth performance is obtained, the better function than the single Geneva' is inherent.

10.5   Motion and Dwell Ratio

Assuming the number of grooves of primary Geneva wheel is N_1, then primary Geneva wheel will turn $1/N_1$ revolutions as the driving crank rotates one .

Similarly, if the secondary Geneva wheel has N_2 grooves, the output Geneva wheel will turn $1/N_2$ revolutions as the secondary crank (i.e. primary Geneva wheel) turns one . As two Geneva wheels connected in series, the output Geneva wheel will turn $1/N_1 N_2$ turns as the primary crank revolves one revolution. The ratio of dwell to motion per cycle

$$R = \frac{1 - \dfrac{1}{N_1 N_2}}{\dfrac{1}{N_1 N_2}} = N_1 N_2 - 1 \tag{10.9}$$

Table 10.2 is the time ratio of dwell to motion. R computed by eq. (10.9) would be much more for single pin driving. R may decrease if more pins are equally distributed on the driving disk (i.e. crank). Thus the output revolutions will be $n_1 n_2 / N_1 N_2$ turns if the number of pins on the primary and secondary disks are n_1, n_2 respectively, see eq. (10.10).

$$R = \frac{1 - \dfrac{n_1 n_2}{N_1 N_2}}{\dfrac{n_1 n_2}{N_1 N_2}} = \frac{N_1 N_2}{n_1 n_2} - 1 \tag{10.10}$$

The number of pins on driving disk can not be chosen freely. It must obey the following rules

$$\frac{N_1}{n_1} = \text{integer}, \quad \frac{N_2}{n_2} = \text{integer}$$

Figures in Table 10.2 are obtained by different numbers of driving pins. $R=0$ is the case that the output Geneva wheel rotates continuously without dwell.

Table 10.2   The Time Ratio of Dwell to Motion

No. of grooves, N		No. of driving pins n		Ratio of dwell to motion, R
Primary Geneva wheel	Secondary Geneva wheel	Primary driving crank	Secondary driving crank	
6	3	1	1	17
		2	1	8
		1	3	5
		3	1	5
		2	3	2
		6	1	2
		3	3	1
		6	3	0
4	4	1	1	15
		1	2	7
		2	1	7
		1	4	3
		2	2	3
		4	1	3
		2	4	1
		4	2	1
		4	4	0

11 Double-Cam Mechanisms

11.1 Path Generator

Generating a prescribed path is the most common application of a double-cam

mechanism for its accuracy. It is accepted by automatic machines greatly.

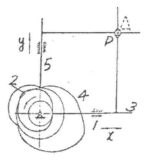

Figure 11.1 Path generating double-cam mechanism

Path generation is usually accomplished by two cams connected   in parallel and

coaxial,   Fig. 11.1. Cams 2 and 4 are integrated with altogether. Completing

the path $y = f(x)$ of point P on the sliders fully depends upon a proper design

of cam profiles.

Now we shall design a double-cam mechanism to accomplish the trace of "A"

graphically. The procedure is as follows (Fig.11.2):

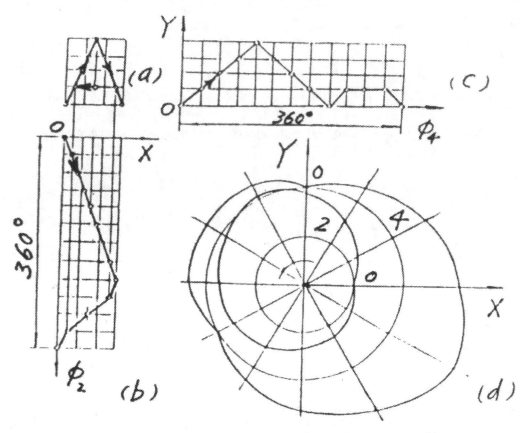

Figure 11.2 Design diagram of trace "A": (a) point path sequence; (b)

$x = f(\phi_2)$;(c) $y = f(\phi_4)$; (d) profiles of cams 2,4.

1. Draw the path A's point sequence, total twelve sections from point to point along the path designated.

2. Draw the curves of the follower displacement versus the angular displacement of the cams, $x = f(\phi_2), y = f(\phi_4)$ respectively

3. The cam profile is obtained according to the graphical method of cam design. The displacement of x is accomplished b link 3, and that of y by link 5. The base circles of cams 2,4 are determined according to spaces provided respectively. Considering the interference and angle of pressure of the cam mechanism are necessary.

11.2 Function Generators

Double-cam mechanisms connected in series can be a function generator. It has the advantage of arbitrarily chosen facility, however, it has the disadvantage of sliding between the contour of themselves. The wear of sliding is a serious problem which may puzzle the designer's determination.

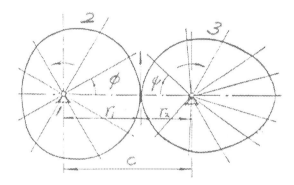

Figure 11.3 Double Cam function generator

Fig.11.3 shows the generator, containing input cam2 and output cam 3. Assume that there is no slip between the contours of the cams. The rotating speed of input cam shaft 2 is constant ω, the output cam shaft 3 is $\dot{\psi}$, then

$$r_1\omega = r_2\dot{\psi}$$

$$\therefore r_2 = \frac{c}{\dot{\psi}/\omega + 1}$$

where the distance between the two shafts is $c = r_1 + r_2$.

Let the transmission ratio $n = 1$ hence

$$\int_0^{2\pi} d\psi = 2\pi \text{ or}$$

$$\int_0^{\frac{2\pi}{\omega}} \dot{\psi} dt = 2\pi$$

The radius r_2 must represent a periodic variation in 2π radius of ψ, or say, $2\pi/\omega$ seconds of t. For instance, if we want to design a pair of cams to accomplish the function

$$\dot\psi = \omega(1+0.5\cos\omega t)$$

Thus, let

$$\int_0^{\frac{2\pi}{\omega}} \dot\psi dt = 2\pi \text{ first, then}$$

$$r_2 = c/(2+0.5\cos\omega t)$$

It may complete a cycle in the interval of $2\pi/\omega$. The problem is how to make the relation of $r_1 = f(\phi), r_2 = f(\psi)$ becomes true. Since

$$r_2 = \frac{c}{2+0.5\cos\omega t}$$

$$r_1 = c - r_2 = c\left(1 - \frac{1}{2+0.5\cos\omega t}\right) = c\left(\frac{1+0.5\cos\omega t}{2+0.5\cos\omega t}\right)$$

where $\omega t = \phi$, hence

$$r_1 = f(\phi) = c\left(\frac{1+0.5\cos\phi}{2+0.5\cos\phi}\right)$$

$$\psi = \int_0^{} \dot\psi dt = \omega\int_0^{}(1+0.5\cos\omega t)dt$$

$$= \omega t + 0.5\sin\omega t$$

$$= \phi + 0.5\sin\phi$$

the series values of r_1, ψ, r_2 can be obtained for every increment of ϕ. Thus the profiles of the two cams shown in Fig. 11.3 are completed.

If a pair of cams set to generate function

$$\dot\psi = \omega(1+\cos\omega t), \text{ then}$$

$$r_1 = c\left(\frac{1+\cos\omega t}{2+\cos\omega t}\right), \text{ should be obeyed. However, this means } r_1 = 0 \text{ when } t = \frac{\pi}{\omega}.$$

It is impossible in fact. Then a differential should be adopted, Fig. 11.4. Let

$$-\omega, \omega(1+K+\cos\omega t) \text{ input to the two cams respectively, then the output will be}$$

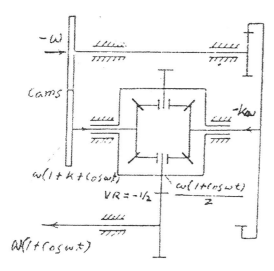

Figure 11.4 Double-cam mechanism with differential connection. K= No. of teeth

of pinion/No. of teeth of ring; No. of teeth of output gear/No. of teeth of

differential housing=1/2

$$\dot{\psi} = \omega\left(1 + K + \cos\omega t\right)$$

where K is the transmission ratio between the pinion and the internal ring gear.

Besides, the transmission ratio of the differential housing to output gear is 2:1.

Then

$$r_1 = \frac{c\left(1 + K + \cos\omega t\right)}{2 + K + \cos\omega t}$$

$$r_2 = \frac{c}{2 + K + \cos\omega t}$$

12　Chained Linkage

12..1　Introduction

The chain drive is different from the rigid member mechanisms described above which is flexible, however. It not only gains wide applications in the transmission field but plays an important partner in composite mechanism also.

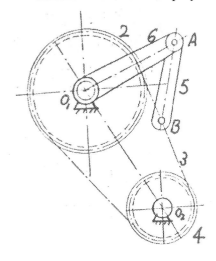

Figure 12.1　Chained linkage (frame not shown)

The chained linkage introduced in this section is just an example of its various kinds. A typical one is shown in Fig.12.1. Members 1,2,3,4 is a chain drive, whereas links 1,3,5,6 is a four-bar linkage. It belongs to the so called reverted mechanism due to its output shaft centerline coaxial with its input shaft's. Energy inputs from the chain sprocket 2 through chain 3 (bypassing sprocket 4) and pin B on chain 3 which forces link 5 to push the output link 6. Helping with the length variations of links 5 and 6, the distance between two sprocket shafts, and radii of sprockets, an output motion spectrum of different kinds of constant speed, various speed, dwell, and reversing are presented. Many reference materials may show us the vast amount of them, however, we have just discussed a little.

12.2 Kinematic Analysis on a Chained Linkage

When the driving sprocket is making a constant speed rotation, all the velocities of the points on the centerline of the chain 3 are equal in magnitude. However, the angular speed of output link 6 is different according to which district the point B belongs to. There are four districts, PQ, QR, RS, and SP. Points P , Q , R , S are the common tangent points of the centerlines of chain to the two sprockets circumferences, see Fig.12.2.

The four-bar linkage becomes an offset slider crank mechanism when the chain member which point B belongs to is in districts PQ, RS; crank rocker mechanism in QR ; a structure with sprocket 2 in SP . The output angular velocities ω_6 in these four districts are respectively as follows

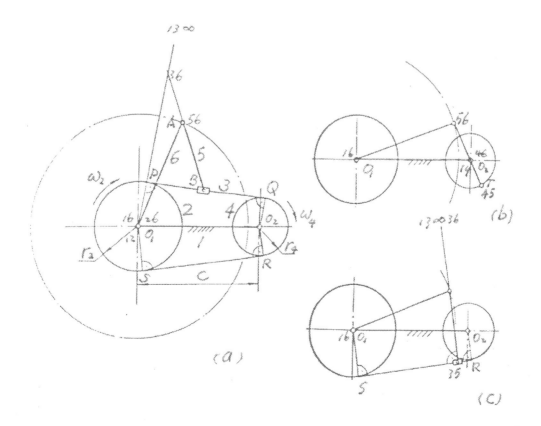

Figure12.2 Chained linkage's kinematic characteristics: (a)

configuration;(b) $o_2 - 46 = 0$ for link 5 coincides with $o_2 - 45$ in district QR ; (c)

$O_1 - 36 = \infty$ for link 5 perpendicular to RS in district RS.

$$
\left.
\begin{aligned}
&\overrightarrow{PQ} : \frac{r_2}{O_1 - 36}\omega_2 \\[2mm]
&\overrightarrow{QR} : \frac{r_2(46 - O_2)}{r_4(46 - O_1)}\omega_2 \\[2mm]
&\overrightarrow{RS} : \frac{r_2}{O_1 - 36}\omega_2 \\[2mm]
&\overrightarrow{SP} : \omega_2
\end{aligned}
\right\}
\tag{12.1}
$$

where r_2, r_4, ω_2 are the radii of sprockets 2 and 4, and the angular velocity of

sprocket 2 respectively. Points 36 and 46 are the instantaneous centers of links 3

and 6, 4 and 6 respectively. Hence $O_1 - 36, O_1 - 46, O_2 - 46$ all are variables.

Any time ω_6 is variable if it contains one of these three factors. When point B

comes to T where link 5 coincides with $O_2 - 45$ which represents a line

belonging to link 4 in QR district, Fig.12.2b, ω_6 is zero   due to $O_2 - 46 = 0$.

When link 5 is perpendicular to RS in RS district, ω_6 is zero also due to

$O_1 - 36 = \infty$, Fig.12.2c. These two configurations take place the first order dwell.

Fig.12.3 shows a typical kinematic characteristics of a chained linkage. The main

parameters are:

Number of teeth of sprockets 2,4(i.e. N_2, N_4) ——38,19 respectively

Diameters of sprockets 4,2 (i.e. d_4, d_2) ---- 115.738mm,230.687mm respectively

Pitch $,p$ ---19.05mm

Number of pitches, N_p ----52

Center distance between shafts ,c ----216.1616mm

Length of link 5, l_5 ----95mm

Length of link 6, l_6 ----190mm

The angular velocity curve is the connection of 32-section's discrete values of the

    chain length L computed by eq.(12.1) with instantaneous center method. The

    angular acceleration curve is obtained from the angular velocity curve by

    graphical differentiation. Since the graphical method is based on the mean value

    of small intervals, the curve is a series of straight lines with different slopes.

    Obviously, it is not so accurate, however, being a quality analysis it is suitable.

 The accurate values of ω_6, ε_6 can be computed by analytical method according to

the equally divided sections of the total length of chain.

Besides, the variable speeds dwells, and reversing characteristics of output caused by

the change of related dimensions of c, l_5, l_6, r_2, r_4 ;all these are omitted here, which

remain to be referred to relevant materials and books.

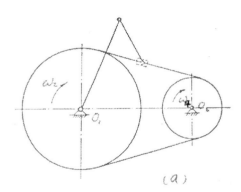

(a)

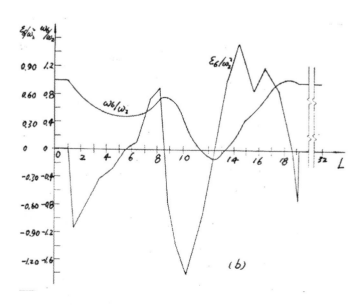

(b)

Figure 12.3 Kinematic characteristic diagram of a chained linkage : (a)

configuration;(b) curve of $\varepsilon_6/\omega_2{}^2$, ω_6/ω_2 versus to chain displacement

Part IV   General Review of Composite Mechanisms

Introduction

Up to now, the analysis and synthesis of the composite mechanisms are not fully and minutely introduced and discussed in this book. Although on real mechanisms, some have been analyzed some have been both analyzed and synthesized; the methods of investigation are quite different. There is no definite rule to obey and to follow. Such a state has existed so long due to the following reasons.

1. The amount of information connected in composite mechanisms is too much. It is very difficult to compose a complete system or to compile an encyclopaedia.

2. The methods used to study are a great deal. A problem might be solved by many methods with different mathematics.

3. No authority could define the composite mechanism strictly. The frontiers of composite mechanisms are boundless.

This book provides many typical examples for analysis and synthesis, mainly in kinematics and with some in statics. There remains many problems to be solved for, and leaves various fields to be exploited A vast amount of research work should be studied continuously.

13   Advantages and Disadvantages of Composite Mechanisms ____

13.1 Geared Linkages

The geared linkage is a kind of composite mechanisms used frequently. The research of it is far and wide. Its presentation is mainly due to the deficiency of planar four-bar linkages. Let us make a comparison between them as follows:

1. Planar four-bar linkages

（ⅰ）as function generator

There are three linear Freudenstein's equations can be applied for three precision positions of the two side links aside the frame on either side. There can be up to four, five, six, and seven equations established for corresponding positions of the two side links, though compatible equations are added for solving. Being nonlinear, the compatible equations make the computation tedious and tiresome.

(ii) as path generator

The number of precision positions can be up to nine. The linear solution is obtained only when the number is not over three.

(iii)as motion generator

The number of precision positions is seven as a limit, whereas the number of linear equations is three still.

(iv) which can produce simple variable speed, dwell, or reversing motion as the drag link mechanism and crank rocker mechanism sometimes do.

(v) whose coupler curves produced by Hrones and Nelson are some more than 7300.

2. Geared linkages

(i) The number of precision positions of the geared linkage is generally more than that of the planar four-bar linkage. As for four-gear five-bar linkage, the number of precision positions for function generation may be up to twelve. The linear equations for linear solution finding is four, better than the planar four-bar linkage still.

(ii)The functions of speed variation, dwell, and reversing are greatly improved, especially on the geared four-bar linkages.

(iii)The types and amount of coupler curves are much more than that of the planar four-bar linkages. Besides, the curves are easier to draw, such as the drawing of two-gear five-bar linkages above.

13.2 Cam-Link Mechanisms

The cam-link mechanism is better (if wear is not considered) than the geared linkage because it can have any number of precision positions, any kind of dwell and reversing motions at the designer's command.

Wear is still the unavoidable drawback of cam mechanisms, which cannot be used in the field of power transmission. On the other hand, cam mechanisms are commonly used in automatic machinery and controlling devices.

As the cam-link mechanism is the main mechanical device for manipulators or robots, the importance is obvious for which are widely spread all over the industrial world.

13.3 Cam-Gear Mechanisms

Generally, cam-gear mechanisms may not only fulfill the motion requirements of cam-link mechanisms but also make the controlling easier for the motions can be strictly carried out by gear mechanisms. For instance, the planetary gear train, rack, and cam composite mechanism comparing with the planetary gear train, rack and Scotch yoke composite mechanism shows the conclusion. Its drawback is still due to wear.

14   Static Force Analysis on Composite Mechanisms

14.1 Introduction

Static force analysis is the foundation of the mechanism research. The composite

mechanism is the very mechanism, therefore on which the static force analysis is

to be studied too. The pressure angle (Appendix L) and the dead center position

(Appendix G) are the key problems of static force analysis.

Apart from the strength, rigidity, friction, and wear; the motion transmission

and the performance of the mechanism are closely related with the pressure

angle and the dead center position of the mechanism itself. The following will

show us minutely the composite mechanism analysis one by one.

14.2 Static Force Analysis on _____

14.2.1   Two-Gear Four-Bar Linkage

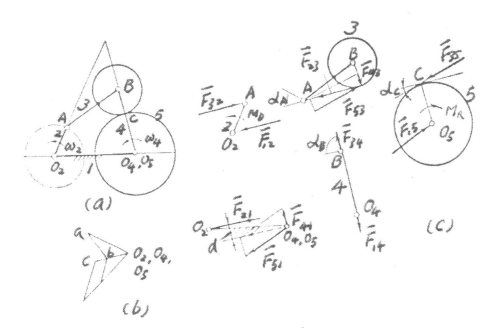

Figure 14.1   Two-gear four-bar linkage:(a) configuration;(b) velocity diagram;(c)

static force analysis, $M_D + M_R = F_{21} \times d, M_D$ ___ driving

moment, M_R ___ working   resistance moment.

Making static force analysis must accompany with the pressure angle discussion.

The two-gear four-bar linkage is a typical composite mechanism, Fig.14.1, however, which has the complexity as well. It should be necessary to define the pressure angle fully before the following discussion for the clarification of some confusion. Supposing a member is subjected in equilibrium by forces of many sources, some are from transmission power, some are from the resistance of other members. The pressure angle defined here is made by the force vector from power source (not from resistance) which acts at the point on the member with the velocity vector at the same point.

As the mechanism in Fig. 14.1, link 1 being the frame(no any velocity at all), link 2 being the power source itself, only links 3,4,5 with pressure angles $\alpha_A, \alpha_B, \alpha_C$ should be discussed.

The α_A is the first to be considered. It is a variable according what position of the mechanism itself is, a spectrum of discrete values of α_A at equally distributed intervals in whole motion cycle is needed for design consideration. The α_B is always 90^0 because the force $\overline{F_{34}}$ from power source is along the link (belonging to two-force member) and velocity $\overline{V}_B$ is always perpendicular to the link which is a member rotating about O_4. No power is transmitted for pressure angle of 90^0. The α_C is the pitch circle pressure angle of involute profile on gear 5. It might be $14.5^0, 15^0$ or 20^0 according to which normal gear system adopted. It has an outstanding advantage of geared linkages as gear is adopted for output member as this example does due to low pressure angle and good transmission quality of gearing.

Another key problem, the dead center positions of mechanisms' shows that no matter how great the input torque is, no output torque can be produced by this mechanism. Therefore the dead center position should be reduced to least for any mechanisms as possible. This book would study this problem more fully. Figs.14.2 and 14.3 are the static force analyses at dead center position for power flowing positively.(from link 2__3__5 conventionally, left to right) and negatively(from link 5__3__2, right to left) respectively.

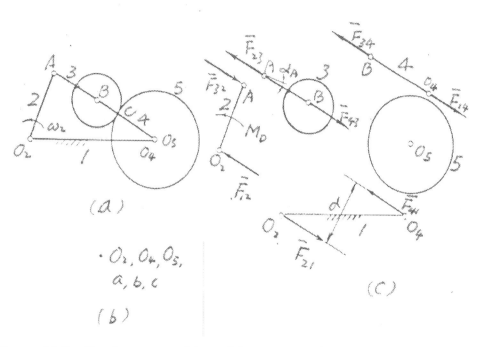

Figure 14.2   Dead center position of the two-gear four-bar linkage while the power

flows positively: (a) configuration; (b) velocity diagram; (c) static force analysis,

$M_D = F_{21} \times d, M_R = 0$.

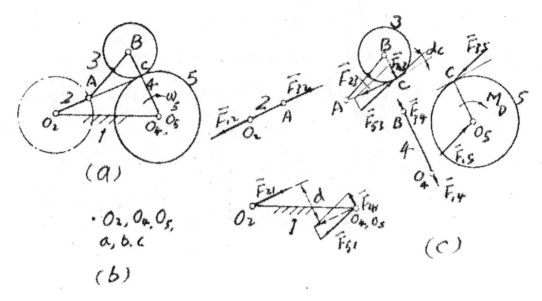

Figure 14.3   Dead center position of the two –gear four-bar linkage while the power

flows negatively: a) configuration; (b) velocity diagram; (c) static force analysis,

$$M_D = F_{21} \times d, M_R = 0$$

All the velocities of the members are equal to zeros in these positions (see velocity

diagrams in related figures). Both the driving torques are absorbed by the

frame. Though the α_A in Fig.14.2 and α_C in Fig. 14.3 are not the real pressure

angles existing for the mechanism is locked, the tendency of the velocities of

corresponding points are obvious, the forces transmitted are real. Therefore

being the virtual pressure angles, they are appeared in the figures still without

significance. So we will not discuss pressure in dead center position any more.

14.2.2   Two- Gear Five- Bar Linkage

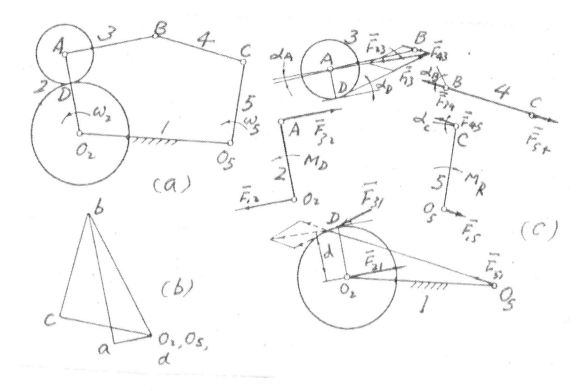

Figure 14.4   Two-gear five-bar linkage: (a) configuration; (b) velocity diagram; (c)

static force analysis, $M_D + M_R = F_{21} \times d$

Assuming power flow in the two-gear five-bar linkage is 2__3__4__5, then pressure

angles $\alpha_A, \alpha_B, \alpha_C$ occur. They are all variables. It is necessary that they should

each have their own spectra. Fig .14.5 is the dead center position where α_A is

shown as   a virtual pressure angle.

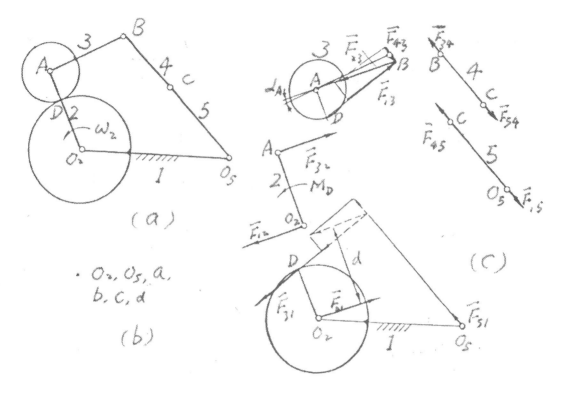

Figure 14.5   Dead center position of the two –gear five-bar linkage while the power flows positively: (a) configuration; (b) velocity diagram; (c) static force analysis, $M_D = F_{21} \times d, M_R = 0$

14.2.3 Four-Gear Five-Bar Linkage

When the power flows in the four-gear five-bar linkage is 2__3__4__5, the static force analysis is much more complex than that of simple one's, see Fig.14.6. The key members of static force   analysis are the two idler gears 6 and 7. Consider the idler gear 6 first for it is simpler than gear 7. According to the equilibrium principle of forces, forces $\overline{F}_{76}, \overline{F}_{36}$ must be intersecting at point E, forces $\overline{F}_{56}, \dot{F}_{46}$ at point F. These two pairs of forces are symmetrical and equal in magnitude, see the free body diagram of gear 6. Through this to gear 5, it would be very clear.   As to gear 7, we obtain forces $\overline{F}_{37}, \overline{F}_{67}$ at point E first, then forces $\overline{F}_{27}', \overline{F}_{17}$ are clear according to the relationships of A, D, E. Furthermore, from the balancing force of the four forces, i.e. $\overline{F}_{27}''$; and the resultant force of

$\overline{F}_{27}{}'$, $\overline{F}_{27}{}''$, i.e. $\overline{F}_{27}$, can be obtained graphically. We need force $\overline{F}_{27}$ for which

is the force of the driving couple M_D. The resultant turning moment on the

frame is $\overline{F}_{71} \times d$ as shown in the figure. There are seven pressure angles

$\alpha_{A3}, \alpha_{A7}, \alpha_{B4}, \alpha_{B6}, \alpha_E, \alpha_F, \alpha_C$. All these are nominal pressure angles of related

members. A spectra of all these angles is needed.

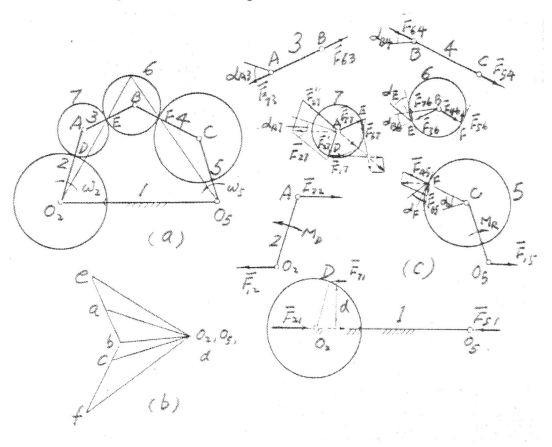

Figure 14.6   Four-gear five-bar linkage: (a) configuration;(b) velocity diagram; (c)

static force analysis, $M_D + M_R = F_{71} \times d$

The more the members of the composite mechanism have, the more the dead

center positions exist. Only one dead center position is presented here for

example, Fig.14.7. The statically indeterminate problem should be noted in this

dead center position. For instance, if the driving torque is applied clock-wise,

the power flow through link 3 and gear 7 will be divided into two path ___ the one along the links ,the other along gears. Though at last they are concurrent at

gear 5, the amount along any path is indeterminate. Besides, the dead center

position is unstable, a slight force nonparallel to links 3,4 or 5 would destroy its

equilibrium. As for his example, there is a recovering action automatically due

to links 3,4, and 5 in tension.

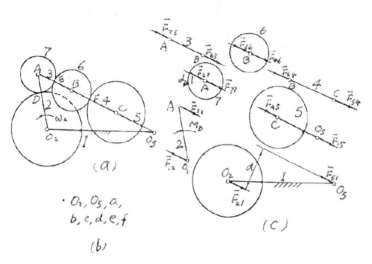

Figure 14.7   Dead center position of the four-gear five-bar linkage while the power

flows positively: (a) configuration; (b) velocity diagram; (c) static force

analysis, $M_D = F_{21} \times d, M_R = 0$

14.2.4   Four-Gear Six-Bar Linkage

Though the four-gear six –bar linkage has one more bar than the four-gear five-bar

linkage, according to the Grubler's criterion the number of members of which

one less instead of one more for all the gears are rigidly attached to

corresponding links in this linkage. So the number of degrees of freedom of this

mechanism

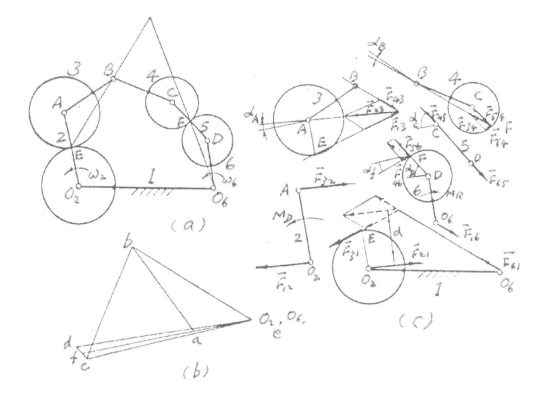

Figure 14.8　Four-gear six-bar linkage: (a) configuration;(b) velocity diagram;(c)

static force analysis, $M_D + M_R = F_{21} \times d$

$F = 3(6-1) - 2 \times 6 - 1 \times 2 = 1$　still.

Only link 5 is a two-force member in static force analysis, see Fig.14.8. The

pressure angles $\alpha_A, \alpha_B, \alpha_C, \alpha_D, \alpha_F$ should be considered effective still. Fig. 14.9

is one of the dead center positions of this mechanism when the power flow is

2__3__4__5__6, the others are neglected

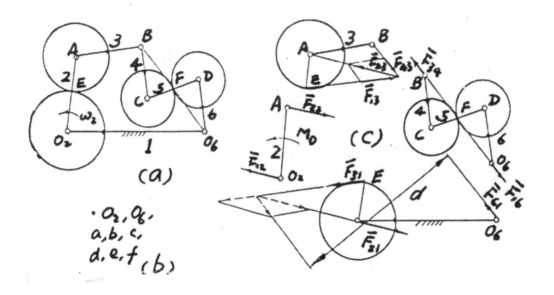

Figure 14.9   Dead center position of the four-gear six-bar linkage while the power

flows positively: (a) configuration;(b) velocity diagram;(c) static force analysis,

$$M_D = F_{61} \times d, M_R = 0$$

14.2.5   Single-Cam Four-Bar Linkage

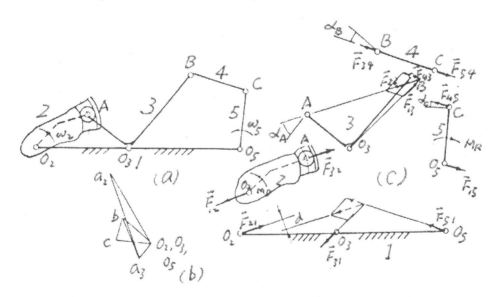

Figure 14.10 Single-cam four-bar linkage: (a) configuration;(b) velocity diagram;(c)

static force analysis, $M_D + M_R = F_{21} \times d$

This single-cam four-bar linkage belongs to the cam-link mechanism connected in series. The static force analysis can be studied by the characteristics of the cam mechanism and linkage separately, so it is simpler than analyzed altogether. Angle α_A is the typical pressure angle belongs to the cam mechanism, whereas α_B, α_C to four-bar linkage. Once the dead center position exists in anyone of the two component mechanisms, the whole mechanism does too. Fig.14.11 is one of the dead center positions of this mechanism

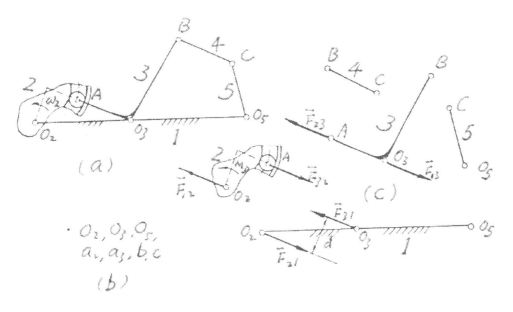

Figure 14.11 Dead center position of the single-cam four-bar linkage while power flows positively;(a) configuration;(b) velocity diagram;(c) static force analysis, $M_D = F_{21} \times d, M_R = 0$.

14.2.6 Single-Cam (Fixed) Four –Bar Linkage

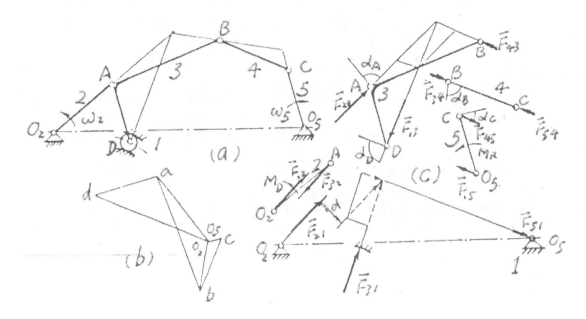

Figure 14.12 Single-am (fixed) four-bar linkage;(a) configuration;(b) velocity

diagram;(c) static force analysis, $M_D + M_R = F_{21} \times d$

The single-cam(fixed) four-bar linkage (Fig.14.12) belongs to the cam-link

mechanism connected in series too. Pressure angles α_A, α_B are to be considered.

Angle α_D keeps 90^0 always neither produces any profit nor consumes energy.

Fig.14.13 is one of the dead center positions of this mechanism.

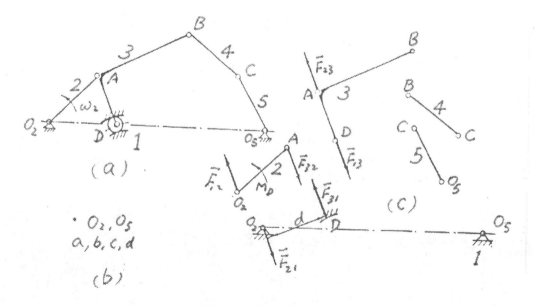

Figure 14 13   Dead center position of the single-cam(fixed) four-bar linkage

while the power flows positively:(a) configuration;(b) velocity diagram;(c) static force analysis,

14.3　Conclusion

1.Static force analysis is the foundation of mechanism mechanics. Pressure angle is an important designation of mechanism performance, which is the by-product of static force analysis and velocity analysis. Discussing on static forces among all the composite mechanisms, of which the geared linkage is the simpler and the camed gear mechanism is the simplest. This text has explained the complex ones.

2.When the dead center position occurs, some two-force members may have an enormous force created if a little bit of force is applied to the point where the virtual pressure angle of value 90^0 becomes true. The pressure angle on this point can be beneficial, as shown in Appendix L, Fig.L..2.,

3.Sometimes the pressure angle of 90^0 might exist on smooth surface or frictionless joint. No energy is consumed or produced. Pressure angle need not be considered in these cases.

4.As pressure angle on every member may vary from $0^0 \rightarrow 90^0$ at any instant, so the key members should be analyzed fully for transmission quality control.

5.The pressure angle must locate on the point of two-force member where the power source inputs. As to three-force and more than three-force members, the pressure angle also exists where the energy is inputing.

6.The statically indeterminate conditions frequently occur at dead center positions, such as on the two-gear four-bar linkage (Fig.14.2) and four-gear five-bar linkage (Fig.14.7) if both the M_D's are reversed. The power flows through two passages. The ratio of the flow amounts between them is indeterminate.

15 An Example for Composite Mechanism Synthesis——A Graphical Synthesis for Output Stroke Extension

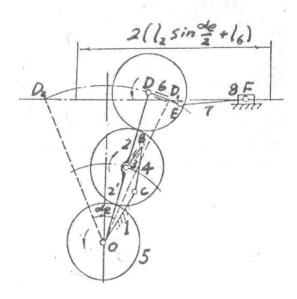

Figure 15.1   Composite mechanism for output stroke extension

Design a geared linkage for the purpose of output stroke extension as shown in Fig.15.1. The basic mechanism is a double rocker mechanism. The prescribed lengths of links 2, 6, and 7 (i.e. l_2, l_6, l_7) are 1 (the unity in this example), 0.2, and 0.6 respectively. The path of the pin F on slider 8 coincides with the horizontal line D_1D_2——the connecting line of the two extreme points of the common joint of links 2 and 6. The stroke of the slider $S = D_1D_2 + 2l_6 = 2(l_2 \sin \frac{\alpha_e}{2} + l_6)$, where

$\alpha_e (= 60^0$, assumed) is the oscillating angle of link 2. The input gear 5 is rotating with a uniform speed. Gears 3 and 6 are rigidly attached to links 3 and 6 respectively. Find out the lengths of links 1, $2'$, 3, 4 (i.e. $l_1, l_{2'}, l_3, l_4$ or OC, OA, AB, BC), the pitch radii of gears 3, 5, and 6 (i.e. r_3, r_5, and r_6).

Solution:
First, compute the number of degrees of freedom of the mechanism
$$F = 3(8-1) - 2 \times 9 - 1 \times 2 = 1$$
Secondly, design the double rocker mechanism (i.e. to do the dimensional synthesis——find the lengths of links 1, $2'$, 3, and 4). Assuming one of the rockers, link $2'$, is prescribed (no matter it is one or one xth of link 2). According to the characteristics of the double rocker mechanism, if link 2' completes a fully

cycle of oscillating motion, link 3 will have revolved a complete revolution. Furthermore, by the nature of differential mechanism, the rotations of gears 5 and 6 will be synchronous at any instant if they are identical gears. The intention of this example is that link 6 must revolve 180^0 for the leading half motion cycle and another 180^0 for the following half cycle. As a matter of fact, gears 3 and 6 are rigidly attached to links 3 and 6 respectively. Hence the only way to meet the requirement is, gears 5, 3, and 6 all are to be the same diameters among them. Whereas $l_{2'}$, must be equal to one half the length of link 2, i.e. $l_{2'} = \dfrac{l_2}{2}$.

Now tabulate the motion of the planetary gear train as follows:

Table 15.1   The leading half cycle——Link 2 moves from right limit position OD_1 to left limit position OD_2

	Arm 2	Gear 5	Gear 3	Gear 6
Motion with arm relative to frame α_e	$+\dfrac{1}{6}$	$+\dfrac{1}{6}$	$+\dfrac{1}{6}$	$+\dfrac{1}{6}$
Motion relative to arm α_r	0	$(+\dfrac{2}{6})$	$(-\dfrac{2}{6})$	$(+\dfrac{2}{6})$
Total motion relative to frame α_a	$+\dfrac{1}{6}$	$+\dfrac{1}{2}$	$(-\dfrac{1}{6})$	$+\dfrac{1}{2}$

Table 15.2   The following half cycle——Link 2 moves from OD_2 to OD_1

	Arm 2	Gear 5	Gear 3	Gear 6
Motion with arm relative to frame α_e	$-\dfrac{1}{6}$	$-\dfrac{1}{6}$	$-\dfrac{1}{6}$	$-\dfrac{1}{6}$
Motion relative to arm α_r	0	$(+\dfrac{4}{6})$	$(-\dfrac{4}{6})$	$(+\dfrac{4}{6})$
Total motion relative to frame α_a	$-\dfrac{1}{6}$	$+\dfrac{1}{2}$	$(-\dfrac{5}{6})$	$+\dfrac{1}{2}$

Table 15.3   One complete cycle——Link 2 oscillates back and forth once

	Arm 2	Gear 5	Gear 3	Gear 6

Total motion relative to frame α_a	0	+1	-1	+1

Notes (a)   The sense of rotation in these table is negative while rotating clockwise, positive counterclockwise.

(b)   The unit of angular displacement is one revolution.

(c)   In these tables, the data in parenthesis are the results of computation.

Observing the tables above, we know that the angular velocity ratios of link 2, gears 5, 3, and 6 in leading half cycle are 1:3:-1:3. It is true at any   instant due to gear train characteristics. The ratios in the following half cycle are -1:3:-5:3, and it is true at any instant too. The total ratios are 0:1:-1:1 respectively in a whole cycle computed average out. Just average out, not instantaneously.

Now we are going to discuss the motion in detail. Though the motion of the leading half is not the same as the following half in quantity, the leading half is to be studied minutely and the following half will be left out for its quality is very similar to that of the leading half.

Let the right limit position is the initial position 1, all the initial position angles

$$\alpha_{e_1}, \alpha_{r_1}, \alpha_{a_1} \quad \text{are}$$

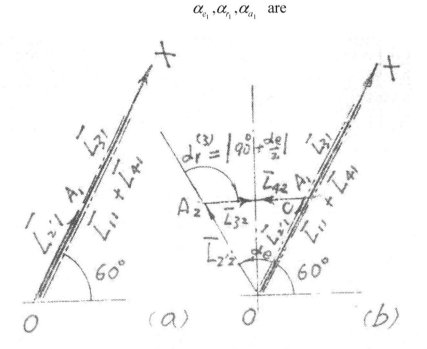

Figure 15.2   The skeleton of a double rocker mechanism: (a) first position, (b) links 1, 2', 3 and 4 in the right (i.e. first) and left limit positions where $\alpha_r^{(3)}$

is the α_r of link 3 when $\alpha_{r_1}^{(3)} = 0$

equal zero, because vector $\overline{OA_1}$ (i.e. link 2') coincides with X-axis Fig. 15.2; the number of the subscripts of the angles are the ordinal numbers of positions.

$$\alpha_{e_1} = 0, \qquad \alpha_{r_1} = 0, \qquad \alpha_{a_1} = 0$$

All the senses of links $1, 2', 3,$ and 4 are the same as X-axis, then

$$\overline{L}_{2'1} + \overline{L}_{31} = \overline{L}_{11} + \overline{L}_{41}$$

Once the $\overline{L}_{2'1}$ is prescribed, the others will be solved. The angular displacements from right limit position to left limit position are

$$\alpha_e = \alpha_{e_2} - \alpha_{e_1} = \alpha_{e_2}$$

$$\alpha_r = \alpha_{r_2} - \alpha_{r_1} = \alpha_{r_2}$$

$$\alpha_a = \alpha_{a_2} - \alpha_{a_1} = \alpha_{a_2}$$

Observing Table 15.1, we know that the α_e's of all the members are the same. Furthermore, the α_e of link $2'$ is the same as α_a of link $2'$, but those of other members are different for their α_r s are different. The total motion of link 3 relative to frame, $\alpha_a^{(3)}$, is $-1/6$ revolutions; the motion relative to arm, $\alpha_r^{(3)}$, is $-1/3$ revolutions. According to the characteristics of double rocker mechanism, links 3 and 4 must coincide with X-axis at the right limit position; whereas at the left limit position they have to join along a straight line horizontally. Therefore, draw a line from A_2 horizontally intersecting X-axis at point C (i.e. the rotating center of link 4) which coincides with point A_1, the terminal A of link $2'$ at its right limit position, Fig.15.2. The length of link 1 is OC. The difference of links 3 and 4 is zero, and the sum of them is A_1A_2, that is

$$l_3 = l_4 = \frac{1}{2} A_1 A_2$$

The above case is just a curious coincidence, because the angular displacement of link 3 relative to link 2, $\alpha_r^{(3)}$, is equal to $|90^0 + \frac{\alpha_e}{2}|$. If $\alpha_r^{(3)}$ is not equal to

$|90^0 + \frac{\alpha_e}{2}|$, point C must be on line OA_1 wandering around A_1, no longer coincident with A_1.

Though the change point never disturbs the operation of the mechanism for the gear mechanism having the unified behavior, it is a pity that restriction is still on line OA_1. If we want point C off the line OA_1, then $\alpha_{r_1}^{(3)}$ must not be equal to zero. We would not change all the related parameters at once, $\alpha_e = 60^0$ still. Let $\alpha_{r_1}^{(3)} > 0$, that is, turn the coincident line of links 3 and 4 counterclockwisely an angle of $\alpha_{r_1}^{(+3)}$ (the positive sign in the right top parenthesis represents the sense of rotating premise). The turning of the coincident line makes the lower boundary line of $\alpha_r^{(3)}$ (i.e. the joined straight line of links 3 and 4) turns also , see Fig.15.3. The shaft center C of link 4 is to be found at C_1. It is to be noted that $\alpha_r^{(3)}$ is a constant if α_e is determined, and it is always equal to $\alpha_{r_2}^{(3)} - \alpha_{r_1}^{(3)}$. When $\alpha_{r_1}^{(3)}$ increases from O to $\alpha_{r_1}^{(+3)}$, then $\alpha_{r_2}^{(3)}$ increases the same amount. Now we know that C_1A_2 is $l_3 + l_4$, C_1A_1 is $l_4 - l_3$. Draw a circle S_{c1A1} with radius C_1A_1, center C_1; another concentric circle S_{C1A2}

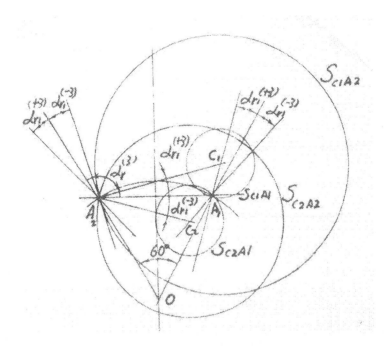

Figure 15.3　The effect of $\alpha_{r_1}^{(3)} \neq 0$ when $\alpha_e = 60^0$

with radius C_1A_2. The space between these two circles is the free movement space of point A. Unfortunately, arc A_2A_1 enters the forbidden region of S_{C1A1} nearby point A_1, the mechanism is being chocked. The prescribed motion will never be completed. Consequently, the boundary line representing the clockwise turned angular displacement $\alpha_{r_1}^{(-3)}$ and the shaft center C_2 of link 4 are to be drawn in Fig.15.3. Circle S_{C2A1} encounters arc A_2A_1 at point A_1 nearby. The same destiny is the mechanism. Through the discussion above, we know that if

$$\alpha_e = 60^0, \qquad \begin{cases} \alpha_{r_1}^{(3)} = 0 & \text{one solution} \\ \\ \alpha_{r_1}^{(3)} \neq 0 & \text{no solution} \end{cases}$$

But what the results are if

$$\alpha_e \neq 60^0, \qquad \begin{cases} \alpha_{r_1}^{(3)} = 0 \\ \\ \alpha_{r_1}^{(3)} \neq 0 \end{cases}$$

Table 15.4 answers us for a variety of α_e values and their corresponding parameters.

Table 15.4   The number of revolutions of link 3 for the leading half cycle

α_e	$+\dfrac{1}{3}$	$+\dfrac{1}{4}$	$+\dfrac{1}{4.5}$	$+\dfrac{1}{5}$	$+\dfrac{1}{6}$	$+\dfrac{1}{7}$	$+\dfrac{1}{8}$	$+\dfrac{1}{9}$	$+\dfrac{1}{10}$
$\alpha_r = n\alpha_e$	$-0.5\times\dfrac{1}{3}$	$-1\times\dfrac{1}{4}$	$-1.25\times\dfrac{1}{4.5}$	$-1.5\times\dfrac{1}{5}$	$-2\times\dfrac{1}{6}$	$-2.5\times\dfrac{1}{7}$	$-3\times\dfrac{1}{8}$	$-3.5\times\dfrac{1}{9}$	$-4\times\dfrac{1}{10}$
β	$+\dfrac{1}{4}$	$+\dfrac{1}{8}$	$+\dfrac{1}{12}$	$+\dfrac{1}{20}$	0	$-\dfrac{1}{28}$	$-\dfrac{1}{16}$	$-\dfrac{1}{12}$	$-\dfrac{1}{10}$

Notes (a) $n = 1 - \dfrac{1}{2\alpha_e}$, $\alpha_r = (1 - \dfrac{1}{2\alpha_e})\alpha_e = \alpha_e - \dfrac{1}{2}$.

(b) $\beta = 90^0 + \dfrac{\alpha_e}{2} - |\alpha_r| = \dfrac{1}{4} + \dfrac{\alpha_e}{2} - (\dfrac{1}{2} - \alpha_e) = -\dfrac{1}{4} + \dfrac{3}{2}\alpha_e$.

(c) α_e must be less than $\dfrac{1}{4}$.

(d) β is the slope of the lower boundary line, its sense is the signal of whether it is above (i.e. +) line A_2A_1, or not (i.e. -).

Not all the α'_es listed above are feasible. As an example, assuming $\alpha'_e = +\dfrac{1}{4}$ rev., the joined straight line A_2B should be parallel to coincident line OA_1, point C

located at infinity, Fig.15.4.

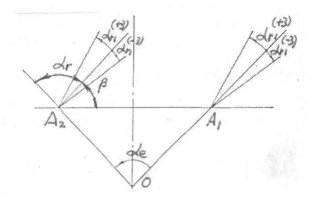

Figure 15.4   Shaft center C at infinity when $\alpha_e = +\dfrac{1}{4}$, $|\alpha_r^{(3)}| = \dfrac{1}{4}$, $\beta = +\dfrac{1}{8}$

No matter how many degrees of $\alpha_{r_1}^{(3)}$ is, clockwise or counterclockwise, there is no solution for the joined straight line revolves the same amount as the coincident line does. As to $\alpha_e = +\dfrac{1}{3}$, the joined straight line A_2B intersects line OA_1 at the other side of AO_2, and the designed double rocker mechanism will be a contradictory product. From the practical point of view, stroke $l_2 \sin \dfrac{\alpha_e}{2}$ is too small when $\alpha_e < \dfrac{1}{10}$. It is of no significance for the extension of stroke.

Let us discuss the case, $\alpha_e = \dfrac{1}{8} rev. = 45^0$, Fig.15.5. Draw lines OA_1 and OA_2 located 45^0 apart symmetrical about the vertical axis on the right and left respectively with O as their center.

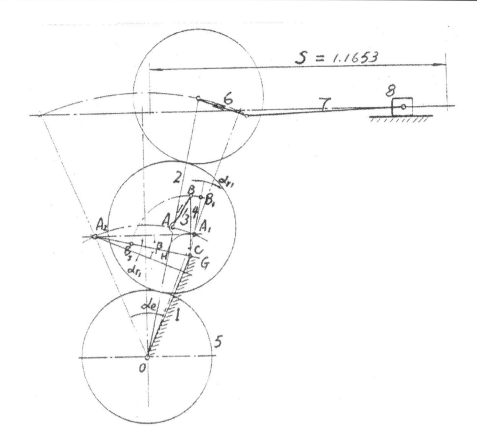

Figure 15.5　The composite mechanism for $\alpha_e = \dfrac{1}{8}$ rev.

Take $\alpha_{r_1} = 10^0$ counterclockwise, though the range can be from 0^0 to

$\beta = \left|-\dfrac{1}{16}rev.\right| = \left|-22.5^0\right|$. (if $\alpha_e = \dfrac{1}{5}rev.$, α_{r_1} must be taken counterclockwisely for

β is positive.) Draw an angle $\beta(= -22.5^0)$, datum A_2A_1, with center A_2, and

the lower boundary line A_2G. Furthermore, measuring an angle equals to α_{r_1}

counterclockwisely, i.e. taking A_2G as datum draw a line intersecting the

extension of line BA_1 (the coincident line of links 3 and 4, obtained from the

revolving of α_{r_1}) at C. Point C is the shaft center of link 4. Since point C is off

line OA_1, i.e. apart from the change point position, the performance of the double

rocker mechanism is modified.

　　Since $A_1C = l_4 - l_3$, $A_2C = l_3 + l_4$, therefore draw an arc with radius CA_1, center

C, intersecting line CA_2 at H. Then A_2H is the difference of A_2C and A_1C, which divided by 2 is l_3. Thus l_4 can be obtained consequently. All the results are as below:

$$l_1 = 0.4149$$
$$l_3 = 0.136$$
$$l_4 = 0.2245$$

$$r_3 = r_5 = r_6 = 0.25$$

$$S = 2(l_2 \sin 22.5^0 + 0.2l_2) = 1.1\ 6\ 5$$

Conclusions

(1) From the above procedure, we know that the satisfactory results will be obtained if the choices of α_e and $\alpha_{r_1}^{(3)}$ are proper. Varying α_e in a possible range, the center C of link 4 will be restricted in the right of line OA_1 by $\alpha_{r_1}^{(3)}$. Solutions can be obtained no matter C is above or under the horizontal line A_2A_1. We may observe it from Figs. 15.3 and 15.4. When C_1 or C_2 is located in the right side of line OA_1 separetely, the path of A can reach point A_1 surely, and never enters the forbidden region.

(2) If we just consider from the view of kinematics, we may take $l_7 = l_6 + l_2(1 - \cos\dfrac{\alpha_e}{2})$. The stroke length will be nearly two folded for the slider passing through the center position of the mechanism while link 6 is just in its middle position, i.e.

$$S = 2[l_2 \sin\frac{\alpha_e}{2} + 2l_6 + l_2(1 - \cos\frac{\alpha_e}{2})]$$

(3) As stated above, gear 3 must be equal to gear 5 because link 2 oscillates back and forth once in a full cycle, gear 5 must rotate a positive revolution, then gear 3 revolves one revolution due to its attaching to link 3, though the sense is negative. In addition, gear 6 (of course including link 6) may change its dimension by relationships as follows:

Table 15.5   Relationships between r_6/r_5 and T/T_0

r_6/r_5	...	$1/7$	$1/5$	$1/3$	$1/1$	$3/1$	$5/1$	$7/1$	......

T/T_0	...	$1/1$	$1/1$	$1/1$	$1/1$	$3/1$	$5/1$	$7/1$	......

Note   "T_0" is the standard unity period of the motion cycle when $(r_6/r_5)=1$,

"T" is the period of the motion cycle of the others.

(4) Now we can draw curves with period as abscissa and stroke as ordinate, say

$$\frac{r_6}{r_5}=\frac{1}{3},\frac{3}{1}, \text{ and}$$

$\frac{1}{1}$ respectively.   The   parameters   prescribed   are

$l_2,\ l_6(=0.2l_2),\alpha_e(=60^0)\ \alpha_{r_1}^{(3)}(=0),r_3(\equiv r_5)$. Let link 7 be a Scotch yoke (i.e.

rigidly attached to slider 8 as an integral; length, infinity; motion as slider;

superposition of two simple harmonic motions——the one is the horizontal

projection of the motion of joint D (the origin of the motion of the moving

translation coordinates), the other is the pin O_1 relative to the moving coordinate.

Table 15.6 and Figure 15.6 are provided for reference. This time, this book has not

shown these curves which would remain to be completed by readers. It is significant

and useful to do through practice by oneself.

Table 15.6   The variations of ratio among the gears

r_6/r_5	$1/3$	$3/1$	$1/1$
$r_6:2r_3:r_5$	1:6:3	3:2:1	1:2:1

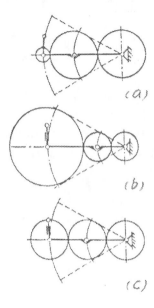

Figure 15.6   Ratios among $r_6, 2r_3, r_5$:

(a)   1:6:3;   (b)   3:2:1;   (c)   1:2:1

From the description above, according to different ratios of r_6 to r_5 and the lengths l_6 and l_7, a spectrum of various characteristics will be presented. The performance of the double rocker mechanism is greatly improved by using the gear trains from the kinematic point of view.

16 Equivalent Mechanisms of a Geared Five-Bar Linkages

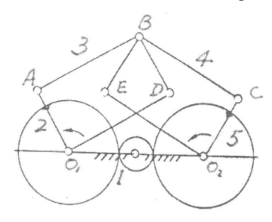

Figure 16.1 Equivalent mechanism of a geared five-bar linkage

If links 2 and 5 shown in Fig. 16.1 are tightly attached to gears 2 and 5 respectively, an idle gear is added between the two gears, the angular velocities ω_2 , ω_5 will be equal in magnitude and sense for their pitch diameter are equal to each other. Let us add links BD, BE, DO_1, EO_2; and let

quadrilaterals BDO_1A , BEO_2C be parallelogram mechanisms. The angle

between lines AO_1 , CO_2 is always keeping unchanged while gear 2 is making

counterclockwise rotation for gear 5 is rotating in the same direction and magnitude as gear 2 does. Then BED becomes a triangle as angle EBD is constant in fact due to the characteristics of parallelogram, and it is to be a member DE of four linkage

O_1O_2DE . Though B is just a common point of links BC and AB, it becomes a point

on link DE of he equivalent four-bar linkage O_1O_2DE . Its characteristics closely

related to DE.

APPENDIXES

APPENDIX   A   Accessible Regions of the Common Joint of Couplers of a Five-Bar
                                 Linkage

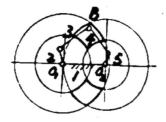

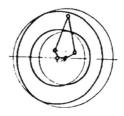

Figure   A.1   Figure   A.2
Two separate accessible regions   Two accessible regions united

Fig.A.1 shows that the pin-connected joint B of the two couplers of a planar five-bar mechanism is just falling into one of the two separate accessible regions, it will be falling into the another one if the mechanism is upside down. The separate accessible regions are two lapped areas generated by the left ring area with $L_3 + L_2$ and $L_3 - L_2$ as its radii, O_1 as center; the right ring area with $L_4 + L_5$ and $L_4 - L_5$ as its radii, O_2 as center respectively.

The two lapped areas can be linked together which is shown in Fig.A.2. It is the result of a suitable design of relative lengths of links.

APPENDIX B   The Path and the Motion Generation (Trace of a Moving Particle and Rigid Body Guidance) of the Planar Four-Bar Linkage

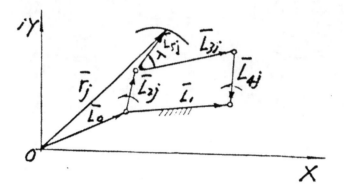

Figure B.1   The Path and the motion generation of a planar four-bar linkage in *j*th position

The two vector equations of closure written for the *j*th positions are as follows:

$$\overline{L}_0 + \overline{L}_{2j} + \overline{L}_{5j} = \overline{r}_j \tag{B.1}$$

$$\overline{L}_{2j} + \overline{L}_{3j} + \overline{L}_{4j} = \overline{L}_{1j} = \overline{L}_1 \tag{B.2}$$

The synthesis equations with three finitely separated precision positions (i.e. precison points) may be represented in matrix as follows:

$$
\begin{bmatrix}
0 & 1 & 0 & 0 & 1 \\
0 & e^{i\phi_{12}} & 0 & 0 & e^{i\gamma_{12}} \\
0 & e^{i\phi_{13}} & 0 & 0 & e^{i\gamma_{13}} \\
-1 & 1 & 1 & 1 & 0 \\
-1 & e^{i\varphi_{12}} & e^{i\gamma_{12}} & e^{i\psi_{12}} & 0 \\
-1 & e^{i\phi_{13}} & e^{i\gamma_{13}} & e^{i\psi_{13}} & 0
\end{bmatrix}
\begin{bmatrix}
\overline{L}_1 \\
\overline{L}_{21} \\
\overline{L}_{31} \\
\overline{L}_{41} \\
\overline{L}_{51}
\end{bmatrix}
=
\begin{bmatrix}
\overline{r}_1 - \overline{L}_0 \\
\overline{r}_2 - \overline{L}_0 \\
\overline{r}_3 - \overline{L}_0 \\
0 \\
0 \\
0
\end{bmatrix}
\tag{B.3}
$$

The path and the motion generation discussion is in following tables.

Table B.1   Path Generation of a Planar Four-Bar Linkage

No.of precision positions	No.of algebraic equations	Unknowns and their numbers	Arbitrarily Chosen reals and their numbers	No.of unkonwns to be solved for
ϕ_{1j}, ψ_{1j} prescribed; linear solution				
1	4	$\overline{L}_0, \overline{L}_1, \overline{L}_{21}, \overline{L}_{31}, \overline{L}_{41}, \overline{L}_{51}$ 12	$\overline{L}_{51}, \lambda$ and three other vectors of the left, 8	4
2	8	Above $+\gamma_{12}$ 13	$\overline{L}_{51}, \lambda, \gamma_{12}$ and one other vector of the left 5	8
3	12	Above $+\gamma_{13}$ 14	γ_{12}, γ_{13} 2	12
ϕ_{1j} prescribed; nonlinear solution				
4	16	Above $+\gamma_{14}, \psi_{12}, \psi_{13}, \psi_{14}$ 18	γ_{12}, γ_{13} 2	16
5	20	Above $+\gamma_{15}, \psi_{15};$ 20	0	20
Nothing prescribed, non-linear solution				
6	24	Above $+\gamma_{16}, \psi_{16}, \phi_{12},$	$\gamma_{12}, \gamma_{13}, \gamma_{14}$	

		$\phi_{13},\phi_{14},\phi_{15},\phi_{16}$ 27		3	24
7	28	Above $+\gamma_{17},\psi_{17},\phi_{17}$ 30	$\gamma_{12},\ \gamma_{13}$ 2	28
8	32	Above $+\gamma_{18},\psi_{18},\phi_{18}$ 33	γ_{12} 1	32
9	36	Above $+\gamma_{19},\psi_{19},\phi_{19}$ 36	 o	36

Notes (a) $\bar{r}_j$ prescribed;

(b) It should be considered whether there are solutions though the parameters are being chosen arbitrarily.

Table B.2   Motion Generation of a Planar Four-Bar Linkage

No. of precision positions	No. of algebraic equations	Unknowns and their numbers	Arbitrarily chosen reals and their numbers	No. of unknowns to be solved for
ϕ_{1j},ψ_{1j} prescribed; linear solution				
1	4	$\bar{L}_0,\bar{L}_1,\bar{L}_{21},\bar{L}_{31},$ $\bar{L}_{41},\bar{L}_{51}$ 12	$\bar{L}_{51},\lambda$ and three other vectors of the left 8	4
2	8	Above 12	L_{51},λ and one vector of the left 4	8
3	12	Above 12	0	12
Nothing prescribed; nonlinear solution				
4	16	Above$+\phi_{12},\phi_{13}$, $\phi_{14},\psi_{12},\psi_{13},\psi_{14}$ 18	L_{51},λ 2	16
5	20	Above$+\phi_{15},\psi_{15}$ 20	0	20

Notes (a) Besides $\bar{r}_j,\bar{r}_{1j}$ prescribed also according to motion requirement;

(b) It should be considered whether there are solutions , though the parameters are being chosen arbitrarily.

APPENDIX C   Linear Equations from Matrix Origin

For the convenience of computation, equation (1.13) can be expanded as follows:

$$
\begin{bmatrix}
1+i0.0 & 1+i0.0 & 1+i0.0 & 1+i0.0 \\
\cos\phi_{12}+i\sin\phi_{12} & \cos\gamma_{12}+i\sin\gamma_{12} & \cos\mu_{12}+i\sin\mu_{12} & \cos\psi_{12}+i\sin\psi_{2} \\
\cos\phi_{13}+i\sin\phi_{13} & \cos\gamma_{13}+i\sin\gamma_{13} & \cos\mu_{13}+i\sin\mu_{13} & \cos\psi_{13}+i\sin\psi_{13} \\
\cos\phi_{14}+i\sin\phi_{14} & \cos\gamma_{14}+i\sin\gamma_{14} & \cos\mu_{14}+i\sin\mu_{14} & \cos\psi_{14}+i\sin\psi_{14}
\end{bmatrix}
\begin{bmatrix}
L_{21x}+iL_{21y} \\
L_{31x}+iL_{31y} \\
L_{41x}+iL_{41y} \\
L_{51x}+iL_{51y}
\end{bmatrix}
$$

$$
=
\begin{bmatrix}
1+i0.0 \\
1+i0.0 \\
1+i0.0 \\
1+i0.0
\end{bmatrix}
\tag{C.1}
$$

Eight linear algebraic equations can be instead of the above matrix as follows:

$$L_{21x}+L_{31x}+L_{41x}+L_{51x}=1$$

$$L_{21y}+L_{31y}+L_{41y}+L_{51y}=0$$

$$L_{21x}\cos\phi_{12}+L_{31x}\cos\gamma_{12}+L_{41x}\cos\mu_{12}+L_{51x}\cos\psi_{12}-L_{21y}\sin\phi_{12}$$

$$-L_{31y}\sin\gamma_{12}-L_{41y}\sin\mu_{12}-L_{51y}\sin\psi_{12}=1$$

$$L_{21x}\sin\phi_{12}+L_{31x}\sin\gamma_{12}+L_{41x}\sin\mu_{12}+L_{51x}\sin\psi_{12}+L_{21y}\cos\phi_{12}$$

$$+L_{31y}\cos\gamma_{12}+L_{41y}\cos\mu_{12}+L_{51y}\cos\psi_{12}=0$$

$$L_{21x}\cos\phi_{13}+L_{31x}\cos\gamma_{13}+L_{41x}\cos\mu_{13}+L_{51x}\cos\psi_{13}-L_{21y}\sin\phi_{13}$$

$$-L_{31y}\sin\gamma_{13}-L_{41y}\sin\mu_{13}-L_{51y}\sin\psi_{13}=1$$

$$L_{21x} \sin \phi_{13} + L_{31x} \sin \gamma_{13} + L_{41x} \sin \mu_{13} + L_{51x} \sin \psi_{13}$$
$$+ L_{21y} \cos \phi_{13} + L_{31y} \cos \gamma_{13} + L_{41y} \cos \mu_{13} + L_{51y} \cos \psi_{13} = 0$$
$$L_{21x} \cos \phi_{14} + L_{31x} \cos \gamma_{14} + L_{41x} \cos \mu_{14} + L_{51x} \cos \psi_{14}$$
$$- L_{21y} \sin \phi_{14} - L_{31y} \sin \gamma_{14} - L_{41y} \sin \mu_{14} - L_{51y} \sin \psi_{14} = 1$$
$$L_{21x} \sin \phi_{14} + L_{31x} \sin \gamma_{4} + L_{41x} \sin \mu_{14} + L_{51x} \sin \psi_{14}$$
$$+ L_{21y} \cos \phi_{14} + L_{31y} \cos \gamma_{14} + L_{41y} \cos \mu_{14} + L_{51y} \cos \psi_{14} = 0$$

$$(C.2)$$

Also, it can be expressed by matrix like this

$$
\begin{bmatrix}
1 & 1 & 1 & 1 & 0 & 0 & 0 & 0 \\
0 & 0 & 0 & 0 & 1 & 1 & 1 & 1 \\
\cos\phi_{12} & \cos\gamma_{12} & \cos\mu_{12} & \cos\psi_{12} & -\sin\phi_{12} & -\sin\gamma_{12} & -\sin\mu_{12} & -\sin\psi_{12} \\
\sin\phi_{12} & \sin\gamma_{12} & \sin\mu_{12} & \sin\psi_{12} & \cos\phi_{12} & \cos\gamma_{12} & \cos\mu_{12} & \cos\psi_{12} \\
\cos\phi_{13} & \cos\gamma_{13} & \cos\mu_{13} & \cos\psi_{13} & -\sin\phi_{13} & -\sin\gamma_{13} & -\sin\mu_{13} & -\sin\psi_{13} \\
\sin\phi_{13} & \sin\gamma_{13} & \sin\mu_{13} & \sin\psi_{13} & \cos\phi_{13} & \cos\gamma_{13} & \cos\mu_{13} & \cos\psi_{13} \\
\cos\phi_{14} & \cos\gamma_{14} & \cos\mu_{14} & \cos\psi_{14} & -\sin\phi_{14} & -\sin\gamma_{14} & -\sin\mu_{14} & -\sin\psi_{14} \\
\sin\phi_{14} & \sin\gamma_{14} & \sin\mu_{14} & \sin\psi_{14} & \cos\phi_{14} & \cos\gamma_{14} & \cos\mu_{14} & \cos\psi_{14}
\end{bmatrix}
\begin{bmatrix}
L_{21x} \\ L_{31x} \\ L_{41x} \\ L_{51x} \\ L_{21x} \\ L_{31x} \\ L_{41x} \\ L_{51x}
\end{bmatrix}
$$

$$
=
\begin{bmatrix}
1 \\ 0 \\ 1 \\ 0 \\ 1 \\ 0 \\ 1 \\ 0
\end{bmatrix}
$$

$$(C.3)$$

APPENDIX   D   Design a Planar Four-Bar Linkage for Four Prescribed Positions —An Example for solving non-linear algebraic equations

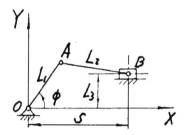

Figure   D.1   An off-set slider crank mechanism

For planar four-bar linkages, when the number of prescribed (or precision) position is over three, say four, then a parameter (usually being the initial angle of output) from the prescribed ones must be picked out as an unknown to be solved for. Four equations can be written down for four prescribed positions, however, as a matter of fact the unknowns will be more than four. Some compatible equations should be added, and nonlinear characteristics would be inherent in. A simple example will explain it clearly for sure as follows:

A slider crank mechanism is shown in Fig.D.1.

$$x_A = L_1 \cos \phi$$
$$y_A = L_1 \sin \phi$$
$$x_B = s$$
$$y_B = L_3$$
$$L_2^2 = (x_B - x_A)^2 + (y_B - y_A)^2$$
$$= (s - L_1 \cos \phi)^2 + (L_3 - L_1 \sin \phi)^2$$

After mathematical treatment, we have

$$2L_1 s \cos \phi + 2L_1 L_3 \sin \phi - (L_1^2 - L_2^2 + L_3^2) = s^2 \qquad (D.1)$$

or $K_1 s \cos \phi + K_2 \sin \phi - K_3 = s^2$

where $K_1 = 2L_1, K_2 = 2L_1 L_3, K_3 = L_1^2 - L_2^2 + L_3^2$

The Freudenstein's three (displacement) equations will be as follows if three prescribed positions are given.

$$K_1 s_1 \cos \phi_1 + K_2 \sin \phi_1 - K_3 = s_1^2$$
$$K_1 s_2 \cos \phi_2 + K_2 \sin \phi_2 - K_3 = s_2^2$$
$$K_1 s_3 \cos \phi_3 + K_2 \sin \phi_3 - K_3 = s_3^2$$

L_1, L_2 and L_3 will be solved for after K_1, K_2, and K_3 are obtained.

As for four prescribed positions, four Freadenstein equations can be listed. The corresponding four unknowns to be solved for are L_1, L_2, L_3 , and either input angle ϕ_1 or output distance s_1 (assume the first position as the initial position).

We adopt s_1 as unknown to be solved for then at jth position

$$s_j = s_1 + s_{1j}$$

where s_{1j} is the displacement from first position to jth position, Eq.(D.1) becomes

$$2L_1(s_1 + s_{1j}) \cos \phi_j + 2L_1 L_3 \sin \phi_j - (L_1^2 - L_2^2 + L_3^2) = (s_1 + s_{1j})^2$$

or

$$2L_1 s_1 \cos \phi_j + 2L_1 s_{1j} \cos \phi_j + 2L_1 L_3 \sin \phi_j - (L_1^2 - L_2^2 + L_3^2 + s_1)^2$$
$$= 2s_1 s_{1j} + s_{1j}^2$$

or

$$K_1 \cos \phi_j + K_2 s_{1j} \cos \phi_j + K_3 \sin \phi_j - K_4 = K_5 s_{1j} + s_{1j}^2 \qquad \text{(D.2)}$$
$$j = 1,2,3,4$$

where $\qquad K_1 = 2L_1 s_1, K_2 = 2L_1$

$$K_3 = 2L_1 L_3, K_4 = L_1^2 - L_2^2 + L_3^2 + s_1^2 , K_5 = 2s_1$$

Eq.(D.2) can be four equations containing five unknowns K_1, K_2, K_7, K_4, and K_5 to be solved for, however, there is a compatible equation with the unknowns, i.e.

$$2K_1 = K_2 K_5$$

(D.3)

or $\qquad\qquad 2K_1 - K_2 K_5 = 0$

Add eq.(D.3) into eq.(D.2) , then solutions can be obtained, However, eq (D.3) is a non-linear equation.

Let $\qquad\qquad K_5 = \lambda$, eq.(D.3) becomes

$$2K_1 - K_2\lambda = 0 \tag{D.4}$$

Eq.(D.2) can be rewritten as

$$K_1\cos\phi_{1j} + K_2 s_{1j}\cos\phi_j + K_3\sin\phi_j - K_4 = \lambda s_{1j} + s_{1j}^{\ 2}$$
$$j = 1,2,3,4 \tag{D.5}$$

Since λ is an unknown, it can not be solved still. Separating the right side of the equation into two parts, we get two equations as follows:

$$l_1\cos\phi_j + l_2 s_{1j}\cos\phi_j + l_3\sin\phi_j - l_4 = s_{1j}$$
$$m_1\cos\phi_j + m_2 s_{1j}\cos\phi_j + m_3\sin\phi_j - m_4 = s_{1j}^{\ 2}$$
$$j = 1,2,3,4$$

whereas

$$K_j = \lambda l_j + m_j, (j = 1,2,3,4) \tag{D.6}$$

substituting eq.(D.6) into eq.(D.3), we have

$$2(\lambda l_1 + m_1) - (\lambda l_2 + m_2)\lambda = 0$$

or

$$l_2\lambda^2 + (m_2 - 2l_1)\lambda - 2m_1 = 0 \tag{D.7}$$

Results will be found when the quadratic equation is being solved. Let us take the following example for explanation.

【Example D.1】

Parameters prescribed:

The displacements of crank angle $\phi_{12},\phi_{13},\phi_{14}$

The displacements of slider s_{12},s_{13},s_{14}

The first position of crank ϕ_1

Unknowns to be solved for L_1, L_2, L_3, and s_1

Solution:

1. calculation

$$\phi_2 = \phi_1 + \phi_{12}, \phi_3 = \phi_1 + \phi_{13}, \phi_4 = \phi_1 + \phi_{14}$$

2. Solving the following two sets linear equations

$$l_1\cos\phi_1 + 0 + l_3\sin\phi_1 - l_4 = 0$$
$$l_1\cos\phi_2 + l_2 s_{12}\cos\phi_2 + l_3\sin\phi_2 - l_4 = s_{12}$$
$$l_1\cos\phi_3 + l_2 s_{13}\cos\phi_3 + l_3\sin\phi_3 - l_4 = s_{13}$$
$$l_1\cos\phi_4 + l_2 s_{14}\cos\phi_4 + l_3\sin\phi_4 - l_4 = s_{14}$$

and

$$m_1 \cos\phi_1 + 0 + m_3 \sin\phi_1 - m_4 = 0$$

$$m_1 \cos\phi_2 + m_2 s_{12} \cos\phi_2 + m_3 \sin\phi_2 - m_4 = s_{12}^{\ 2}$$

$$m_1 \cos\phi_3 + m_2 s_{13} \cos\phi_3 + m_3 \sin\phi_3 - m_4 = s_{13}^{\ 2}$$

$$m_1 \cos\phi_4 + m_2 s_{14} \cos\phi_4 + m_3 \sin\phi_4 - m_4 = s_{14}^{\ 2}$$

Obtain
$$l_j, (j = 1,2,3,4)$$
$$m_j, (j = 1,2,3,4)$$

3. Calculate the discriminant $\Delta = (m_2 - 2l_1)^2 + 8m_1 l_2$

If $\Delta < 0$, no solution

If $\Delta = 0, \lambda = \dfrac{2l_1 - m_2}{2l_2}$

If $\Delta > 0, \lambda = \dfrac{2l_1 - m_2 \pm \sqrt{\Delta}}{2l_2}$

4. Since

$$K_1 = \lambda l_1 + m_1$$
$$K_2 = \lambda l_2 + m_2$$
$$K_3 = \lambda l_3 + m_3$$
$$K_4 = \lambda l_4 + m_4$$

Substituting the values of λ into the above equations, we have,

$$L_1 = \frac{K_2}{2}, s_1 = \frac{\lambda}{2}$$

$$L_3 = \frac{K_3}{K_2}, L_2 = \sqrt{L_1^2 + L_3^2 + s_1^2 - K_4}$$

APPENDIX E   Freudenstein's Displacement Equation

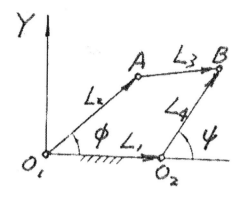

Figure E1   A planar four-bar linkage

A planar four-bar linkage is shown in Fig.E.1.
For A:

$$x_A = L_2 \cos\phi, y_A = L_2 \sin\phi$$

For B:

$$x_B = L_1 + L_4 \cos\psi, y_B = L_4 \sin\psi$$

Assume link AB is an absolute rigid body, hence

$$\left(x_B - x_A\right)^2 + \left(y_B - y_A\right)^2 = L_3^2$$

or

$$\left(L_1 + L_4 \cos\psi - L_2 \cos\phi\right)^2 + \left(L_4 \sin\psi - L_2 \sin\phi\right)^2 = L_3^2$$

Making mathematical treatment, we have

$$D\sin\psi + E\cos\psi = F$$

where

$$D = \sin\phi, E = -\frac{L_1}{L_2} + \cos\phi$$

$$F = -\frac{L_1}{L_4}\cos\phi + \frac{L_1^2 + L_2^2 - L_3^2 + L_4^2}{2L_2L_4}$$

Let

$$\sin\psi = \frac{2\tan\dfrac{\psi}{2}}{1 + \tan\dfrac{\psi}{2}}, \cos\psi = \frac{1 - \tan\dfrac{\psi}{2}}{1 + \tan\dfrac{\psi}{2}}$$

we have

$$\psi = 2\tan^{-1}\frac{D \pm \sqrt{D^2 + E^2 - F^2}}{E + F}$$

These are the two solutions of the output angle ψ by the prescribed input angle ϕ, see Fig.E.2.

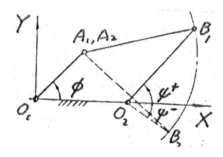

Figure E.2   Two configurations of the four-bar linkage

If we design a four-bar linkage with prescribed positions of input angle ϕ_j and output angle ψ_j, the equations above may be treated as follows:

$$-K_1\cos\phi_j + K_2\cos\psi_j + K_3 = \cos(\phi_j - \psi_j) \qquad (E.1)$$

where

$$K_1 = \frac{L_1}{L_4}$$

$$K_2 = \frac{L_1}{L_2}$$

$$K_3 = \frac{L_1^2 + L_2^2 - L_3^2 + L_4^2}{2L_2 L_4}$$

This is the Freudenstein's equation.

In general, j=1,2,3. Then

$$-K_1\cos\phi_1 + K_2\cos\psi_1 + K_3 = \cos(\phi_1 - \psi_1)$$
$$-K_1\cos\phi_2 + K_2\cos\psi_2 + K_3 = \cos(\phi_2 - \psi_2) \qquad (E.2)$$
$$-K_1\cos\phi_3 + K_2\cos\psi_3 + K_3 = \cos(\phi_3 - \psi_3)$$

There are three unknowns to be solved by these equations. Then the relative lengths of the links can be solved

APPENDIX F   Sylvester's Dialytic Elimination

As for five-point approximation (i.e. five precision position synthesis), there are three unknowns $\lambda_1, \lambda_2, \lambda_3$ hiding in the coefficients of the following polynomials

$$A_1\lambda_4{}^2 + A_2\lambda_4 + A_3 = 0 \tag{F.1}$$

$$B_1\lambda_4{}^2 + B_2\lambda_4 + B_3 = 0 \tag{F.2}$$

$$C_1\lambda_4{}^2 + C_2\lambda_4 + C_3 = 0 \tag{F.3}$$

$$D_1\lambda_4{}^2 + D_2\lambda_4 + D_3 = 0 \tag{F.4}$$

where A_1, B_1, C_1, D_1 are all the functions of known quantities, whereas $A_2, B_2, C_2, D_2, A_3, B_3, C_3 D_3$ are the functions of $\lambda_1, \lambda_{21}, \lambda_3$ (including some known quantities too). To eliminate λ_4 in the above four polynomials, we may use the sylvester's dialytic elimination, Reference [3]. Taking the pairs of eqs. (F.1) and (F.2); (F.1) and (F.3), (F.1) and (F.4) each at a time we obtained three equations containing $\lambda_1, \lambda_2, \lambda_3$. Express these in the following:

$$\begin{vmatrix} A_1 & A_2 & A_3 & 0 \\ 0 & A_1 & A_2 & A_3 \\ B_1 & B_2 & B_3 & 0 \\ 0 & B_1 & B_2 & B_3 \end{vmatrix} = 0 \tag{F.5}$$

$$\begin{vmatrix} A_1 & A_2 & A_3 & 0 \\ 0 & A_1 & A_2 & A_3 \\ C_1 & C_2 & C_3 & 0 \\ O & C_1 & C_2 & C_3 \end{vmatrix} = 0 \tag{F.6}$$

$$\begin{vmatrix} A_1 & A_2 & A_3 & 0 \\ 0 & A_1 & A_2 & A_3 \\ D_1 & D_2 & D_3 & 0 \\ 0 & D_1 & D_2 & D_3 \end{vmatrix} = 0 \tag{F.7}$$

Expanding these as

$$A_1'\lambda_3{}^4 + A_2'\lambda_3{}^3 + A_3'\lambda_3{}^2 + A_4'\lambda_3 + A_5' = 0 \tag{F.8}$$

$$B_1'\lambda_3{}^4 + B_2'\lambda_3{}^3{}_3 + B_3'\lambda_3{}^2 + B_4'\lambda_3 + B_5' = 0 \tag{F.9}$$

$$C_1'\lambda_3{}^4 + C_2'\lambda_3{}^3 + C_3'\lambda_3{}^2 + C_4'\lambda_3 + C_5' = 0 \tag{F.10}$$

where $A_2', A_3', A_4', A_5', B_2', B_3', B_4', B_5', C_2', C_3', C_4', C_5'$ are functions of λ_1, λ_2 and some known quantities. Using Sylvester's dialytic elimination to two equations at a time again, we have two equations of polynomials each containing two unknowns.

$$
\begin{vmatrix}
A_1' & A_2' & A_3' & A_4' & A_5' & 0 & 0 & 0 \\
0 & A_1' & A_2' & A_3' & A_4' & A_5' & 0 & 0 \\
0 & 0 & A_1' & A_2' & A_3' & A_4' & A_5' & 0 \\
0 & 0 & 0 & A_1' & A_2' & A_3' & A_4' & A_5' \\
B_1' & B_2' & B_3' & B_4' & B_5' & 0 & 0 & 0 \\
0 & B_1' & B_2' & B_3' & B_4' & B_5' & 0 & 0 \\
0 & 0 & B_1' & B_2' & B_3' & B_4' & B_5' & 0 \\
0 & 0 & 0 & B_1' & B_2' & B_3' & B_4' & B_5'
\end{vmatrix} = 0 \tag{F.11}
$$

$$
\begin{vmatrix}
A_1' & A_2' & A_3' & A_4' & A_5' & 0 & 0 & 0 \\
0 & A_1' & A_2' & A_3' & A_4' & A_5' & 0 & 0 \\
0 & 0 & A_1' & A_2' & A_3' & A_4' & A_5' & 0 \\
0 & 0 & 0 & A_1' & A_2' & A_3' & A_4' & A_5' \\
C_1' & C_2' & C_3' & C_4' & C_5' & 0 & 0 & 0 \\
0 & C_1' & C_2' & C_3' & C_4' & C_5' & 0 & 0 \\
0 & 0 & C_1' & C_2' & C_3' & C_4' & C_5' & 0 \\
0 & 0 & 0 & C_1' & C_2' & C_3' & C_4' & C_5'
\end{vmatrix} = 0 \tag{F.12}
$$

Expanding these, we have

$$
f_1(\lambda_1, \lambda_2) = a_0 \lambda_2^n + a_1 \lambda_1 \lambda_2^{n-1} + \cdots + a_{n-1} \lambda_1 \lambda_2 + a_n \lambda_1 \tag{F.13}
$$

$$
f_2(\lambda_1, \lambda_2) = b_0 \lambda_2^n + b_1 \lambda_1 \lambda_2^{n-1} + \cdots + b_{n-1} \lambda_1 \lambda_2 + b_n \lambda_1 \tag{F.14}
$$

where $n = 8$; a_0, b_0 are known constants; $a_1 \lambda_1 \cdots, b_1 \lambda_1 \cdots$ are polynomials of λ_1; f_1, f_2 are the polynomials of λ_2. Finally, we have

$$
\begin{vmatrix}
a_0 & a_1 & \cdots & \cdots & a_n & 0 & \cdots & 0 \\
0 & a_0 & a_1 & \cdots & \cdots & a_n & \cdots & 0 \\
\vdots & \vdots & \vdots & \vdots & \vdots & \vdots & \vdots & \vdots \\
0 & \cdots & 0 & a_0 & a_1 & \cdots & \cdots & a_n \\
b_0 & b_1 & \cdots & \cdots & b_n & 0 & \cdots & 0 \\
0 & b_0 & b_1 & \cdots & \cdots & b_n & \cdots & 0 \\
\vdots & \vdots & \vdots & \vdots & \vdots & \vdots & \vdots & \vdots \\
0 & \cdots & 0 & b_0 & b_1 & \cdots & \cdots & b_n
\end{vmatrix} = 0 \tag{F.15}
$$

Expanding and solving this, every real root is the value of λ_1. Substituting back to eqs.(F.13) and (F.14), then at least a common root of λ_2 can be obtained if eq.(F.15) is satisfied. Again substituting λ_1, λ_2 back into eqs. (F.8), (E.9) and (F.10), a common root of λ_3 is obtained. Finally, substituting back $\lambda_1, \lambda_2, \lambda_3$ into eqs. (F.1), (F.2), (F.3), (F.4); λ_4 is obtained.

APPENDIX   G   Dead Center Positions

Introduction

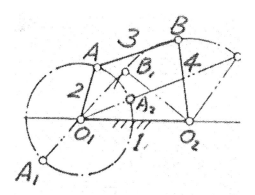

Figure G.1   Dead center positions of a crank rocker mechanism

When the center line of the crank O_1A coincides with the center line of the coupler AB(There are two kinds of coincidence. the one is the crank coincides with the coupler proper, the other is the crank stretching straight coincides with the extension of the center line of coupler.) , i.e. configurations $O_1A_1B_1O_2, O_1A_2B_2O_2$ of the mechanism in Fig.G.1; the rocker is at its limit positions B_1O_2, B_2O_2, the crank may at its dead center positions A_1O_1, A_2O_1 respectively

   It should be noted that the dead center positions occur only if link 4 is the input member. Since link2 cannot turn over a little bit at these positions no matter how great the torque to be applied to input link 4 is, the output torque of link 2 is zero as if it were "dead". On the contrary, a large amount of torque on link 4 will be obtained if a little torque on link 2 is applied.

   No doubt, it is the charm of the dead center position that makes many scholars prostrating themselves at. We shall use the instantaneous center distance method to obtain the dead center positions during the processing for mechanical advantage finding.

Using Instantaneous Center Method as a Theoretical Foundation for Solving the Problem

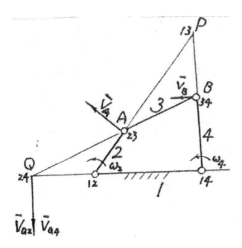

Figure G.2   Instantaneous centers of a crank rocker mechanism

Once all the instantaneous centers of the mechanism in Fig.G.2 are found, consequently the velocities of the related points on the mechanism may be obtained easily. The relevant equations are listed as follows:

$$V_{A2} = (12 - 23)\omega_2$$
$$V_{A3} = (13 - 23)\omega_3$$

where V, ω, and 12-23 (or 13-23) represent linear velocity, angular velocity, and the distance related between the instantaneous centers. The A_2, A_3 are points belong to links 2and 3 respectively and both coincides with the center of joint A where instantaneous center of links 2 and 3 locate.

Since $V_{A2} = V_{A3} = V_A$

Hence

$$\frac{\omega_2}{\omega_3} = \frac{13 - 23}{12 - 23}$$

(G.1)

and $V_{B3} = V_{B4} = V_B$; for the same reason as A_2, A_3's.

$$\frac{\omega_3}{\omega_4} = \frac{14 - 34}{13 - 34}$$

(G.2)

$V_{Q2} = V_{Q4} \neq V_Q$; the reason of being not equality will be proved in Appendix K.

$$\frac{\omega_2}{\omega_4} = \frac{14 - 24}{12 - 24}$$

(G.3)

If the power flows from link 2 to 4 via 3 (i.e. 2__3__4), turning moment is labeled as M , angular velocity as ω , then

$$M_2\omega_2 = M_3\omega_3 = M_4\omega_4$$

(G.4)

Th mechanical advantage

$$MA = \frac{M_4}{M_3} \times \frac{M_3}{M_2} = \frac{\omega_3}{\omega_4} \times \frac{\omega_2}{\omega_3} \qquad (G.5)$$

$$\text{or } MA = \frac{M_4}{M_2} = \frac{\omega_2}{\omega_4}$$

$$(G.6)$$

Substituting eqs. (G.1) and (G.2) into eqs.(G.5); eq. (G.3) into eq.(G.6), we have

$$MA = \frac{13-23}{12-23} \times \frac{14-34}{13-34} \qquad (G.7)$$

$$\text{and } MA = \frac{14-24}{12-24}$$

$$(G.8)$$

If the power flow is reversed, i.e. 4__3__2, then

$$MA = \frac{12-23}{13-23} \times \frac{13-34}{14-34} \qquad (G.9)$$

$$\text{and } MA = \frac{12-24}{14-24}$$

$$(G.10)$$

Eqs.(G.9) and (G.10) are just the reciprocals of eqs.(G.7) and (G.8) respectively. In this case, let link2 joins link 3 stretching on a straight line, then a dead center position appears if power flows from

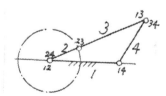

Figure G.3   One of the dead center positions of the crank rocker mechanism

4__3__2,Fig. G.3. Since instantaneous center pairs 13-34 and 12-24 are both coincident points respectively, *MA*'s are equal to zeros from eqs.(G.9) and(G.10.). Whereas the output torque is zero, a dead center position will occur. It should be noted that the denominator must not be zero in *MA* equation used for finding dead center positions, because it may either cause indetermination (i.e.0/0) or impossibility (i.e. constant/0). There is another dead center position when link 2 coincident with link 3 proper as link4 being in its left limit position.

Dead Center Position of Composite Mechanisms

(1)   Two-Gear Four-bar Linkage
 i. Direction of power flow:2__3__5

$$MA = \frac{13-23}{12-23} \times \frac{15-35}{13-3}$$

By the method above, we know that 15-35 cannot be zero, whereas 13-23 is possible. Let 15-35 intersect 12-23 at point 23 give 13, the dead center position occurs as shown in Fig.G.4b.

ii. Direction of power flow: 5__3__2

$$MA = \frac{13-35}{15-35} \times \frac{12-23}{13-23}$$

Connsidering 12-23 cannot be zero, 13-35 is possible. Let 12-23 intersect 15-35 at point 35, the dead center position is obtained, Fig. G.4c.

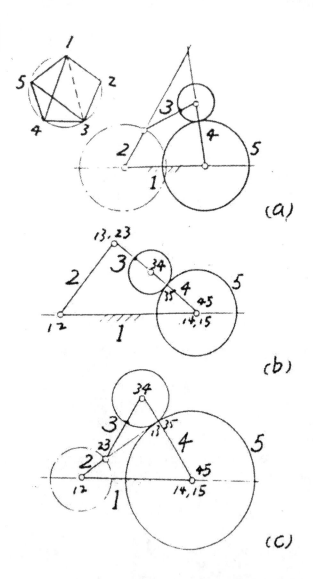

Figure G.4   Two-gear four-bar linkage: (a) configuration; (b) and (c) dead center positions

(2) Two Gear Five-Bar Linkage

i. Direction of power flow: 2—3—4—5

$$MA = \frac{13-23}{12-23} \times \frac{14-34}{13-34} \times \frac{15-45}{14-45}$$

Observing Fig.G.5, we know that both 13-23 and 15-45 cannot be zero, only 14-34 has the possibility . Let 13-34 intersect with link 5 containing 15-45 at point 34 obtaining 14, thus link 5 must be in line with line 4.The dead center position occurs as shown in Fig. G.5b.

ii. Direction of power flow: 5__4__3__2

$$MA = \frac{14-45}{15-45} \times \frac{13-34}{14-34} \times \frac{12-23}{13-23}$$

As 12-23 cannot be zero, whereas both 14-45 and13-34 have the possibility. Fig.G.5c is the dead center position configuration when 14-45 is zero. Fig.G.5d is the dead center position while both 13-34 and 14-4 are equal to zeros, simultaneously it is the special example of Fig.G.5c where geared link 3 only leaves gear proper.

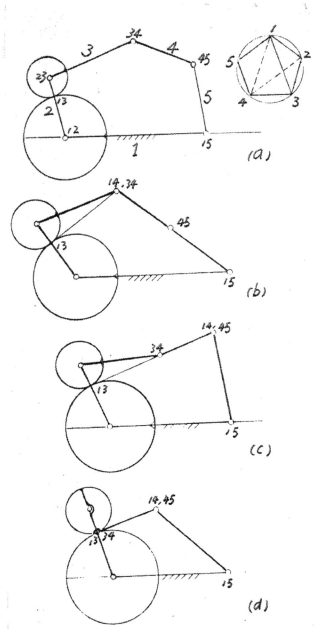

Figure G.5   Two-gear five-bar linkage:(a) configuration;(b),(c), and (d) dead center positions

(3) Four-Gear Five-Bar Linkage

Since both gears 6 and 7 are idler gears, therefore

i. The direction of power flow:2__3__4__5

$$MA = \frac{13-23}{12-23} \times \frac{14-34}{13-34} \times \frac{15-45}{14-45}$$

Find out the instantaneous centers as many as possible. We know that 15-45 cannot be zero, whereas 13-23 and 14-34 have the possibility. First, take aim at 13-23, observe circle diagram at upper right corner in Fig.G6. For finding 13, we must find out12-23 and 16-36; for finding these we must find out 16 first. It is obtained by the intersection of 17-76 and 15-56. Let link 4 line up link 5, then with link 3 afterwards,13 coincides with 23 as shown in Fig.G.6b.

There is another way to define the zero value of 13-23, see Fig.G.6c. Let 15-56 intersect with 17-76 at point 76, 16 being obtained; then let 13 (the intersection of 12-23 and 16-36) coincide with 23.

Secondly, take aim at14-34. The key problem is to find out 14. Whereas 14 is obtained by the intersection of lines 15-45 (belongs to link 5) and 16-46 (46,36, and34 are all coincident points);i.e. links 5 and 4 ,links 3 and 2 are both lined up separately.

Unfortunately, 13 coincides with 34 also; i.e. 13-34=0, making $MA = \dfrac{0}{0}$. Dead center position will not exist for indetermination, Fig.G.6d.

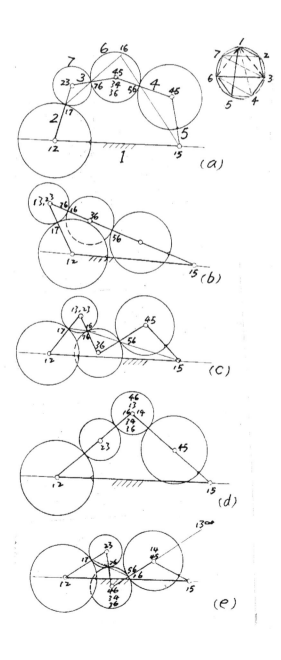

Figure G.6 Four-gear five-bar linkage:(a) configuration;(b),(c) dead center positions;(d) indeterminate position;(e) dead center position

ii. Direction of power flow: 5__4__3__2

$$MA = \frac{14-45}{15-45} \times \frac{13-34}{14-34} \times \frac{12-23}{13-23}$$

Considering 12-23 cannot be zero, but 13-34 can. Simultaneously, 14-34 is zero as proved above. The *MA* becomes indeterminate. So the dead center position does not exist.

As to 14-45, finding 45 is not necessary for it appears naturally. Whereas 14 may be obtained by the intersection of 13-34 and 15-45. The crux of the matter is 13. The 13 is the intersecting point of 12-23 and 16-36. Furthermore, 16 is the crux of the most. The coincidence of 16 and 56 is easy to verify, for the intersection of 17-76 and 15-56 should be 16, which is just on 56 only if link 2 is parallel with link 4. For the same reason, the coincidence of 14 and 45 can occur for 13 is at infinity,Fig.G.6e.

(4) Four-Gear Six-Bar Linkage
i. Direction of power flow: 2__3__4__5__6
$$MA = \frac{13-23}{12-23} \times \frac{14-34}{13-34} \times \frac{15-45}{14-45} \times \frac{16-56}{15-56}$$
Both 13-23 and 16-56 cannot be equal to zeros, whereas 14-34 and 15-45 are possible. Consider 14-34 first. We know that it can be true only if line 16-46 passes through point 34 ,Fig.G.7b and c. Then consider 15-45. It can be zero, and 14-45 is zero simultaneously. There would be an indeterminate position, Fig. G.7d.

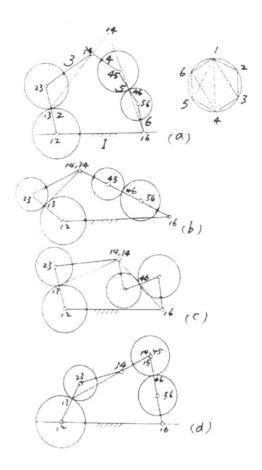

Figure G.7 Four-gear six-bar linkage: (a) configuration;(b) and (c) dead center positions; (d) indeterminate position

ii. Direction of power flow: 6__5__4__3__2
Since the discussion and the procedure are similar to above, so they are neglected.

(5) Cam-Link Mechanism
i. Direction of power flow: 2__3__4__5

$$MA = \frac{13-23}{12-23} \times \frac{14-34}{13-34} \times \frac{15-45}{14-45}$$

The 15-45 cannot be zero whereas both 13-23 and 14-34 can separately, Fig. G.8b and c.. Furthermore, they can combine to be double dead center positions.

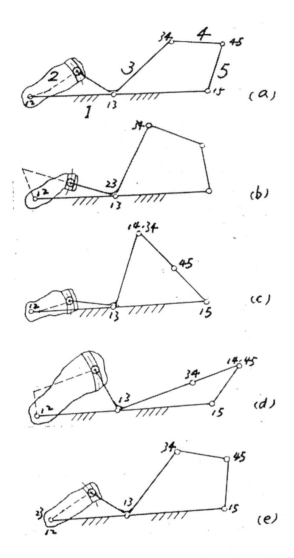

Figure G.8 Cam-link mechanism: (a) configuration;(b) ,(c),(d),(e) dead center positions

ii. Direction of power flow: 5__4__3__2

$$MA = \frac{14-45}{15-45} \times \frac{13-34}{14-34} \times \frac{12-23}{13-23}$$

The 13-34 cannot be zero, whereas both 14-45 and12-23 can, as shown in Fig.G8d and e. They can be double dead center positions too.

Closing Words

(1)   The discussion on the dead center positions of composite mechanisms is necessary, because some positions seem as dead centers are not really in fact and vice versa. The material provided here is just a possibility. Any change of the relative link lengths may lead to impossibility.

(2)   The finding of instantaneous centers may be very difficult due to the increasing amount of link number. So it would be an art to plan for the procedure.

(3)   There might be three instantaneous centers meeting at a point in composite mechanisms, i.e. the special case of Kennedy's theorem presents.

(4)   The theoretical base of dead center position finding is to let mechanical advantage equal zero, namely its numerator equal to zero and the denominator equal to nonzero real number (the premise of this approach is a fractional expression of instantaneous center distances).

(5)   The dead center position should accompany with static force analysis for verification.

(6)   The other composite mechanisms such as cam-gear mechanisms ,Geneva mechanisms ,chain and linkage combinations, and so on are simpler than those discussed above, therefore they are to be neglected.

APPENDIX H   Relationship Between the Input and Output of Crank Rocker Mechanism

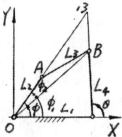

Figure H1   A typical crank rocker mechanism showing the relationship between its input and output links

A crank rocker mechanism is shown in Fig.H.1. Let $\phi = \phi_1 + \phi_2$, then

$$\tan\phi_1 = \frac{L_4 \sin\theta}{L_1 + L_4 \cos\theta}$$

$$\cos\phi_2 = \frac{k + 2L_1 L_4 \cos\theta}{2mL_2}$$

$$\phi = \tan^{-1}\frac{L_4 \sin\theta}{L_1 + L_4 \cos\theta} + \cos^{-1}\frac{k + 2L_1 L_4 \cos\theta}{2mL_2}$$

where $m^2 = L_1^2 + L_4^2 + 2L_1 L_4 \cos\theta$

$$k = L_1^2 + L_2^2 - L_3^2 + L_4^2$$

Besides, according to the instantaneous center 13 and the angular velocities $\dot{\phi}, \omega_3, \dot{\theta}$ of links 2,3,4 respectively, we have

$$(A - 13)\omega_3 = L_2\dot{\phi}$$

$$(B - 1\grave{3})\omega_3 = L_4\dot{\theta}$$

whereas

$$A - 13 = \frac{L_1 \sin\theta}{\sin(\theta - \phi)} - L_2$$

$$B - 13 = \frac{L_1 \sin\phi}{\sin|\theta - \phi|} - L_4$$

After mathematical treatment ,we have

$$\phi = \int\frac{L_4(L_1 \sin\theta - L_2 \sin(\theta - \phi))}{L_2(L_1 \sin\phi - L_4 \sin(\theta - \phi))}d\theta$$

APPENDIX I   Graphical Synthesis on Double Rocker Mechanism for Its Output Oscillating Angle Extension

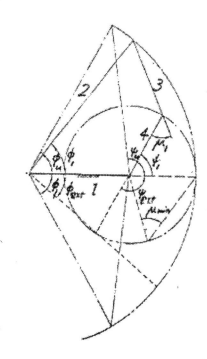

Figure I.1 Double rocker mechanism with large output oscillating angle

A typical double rocker mechanism for large output oscillating angle is shown in Fig.I.1. As the driving link 2 in its top limit position, makes a positive limit angle of ϕ_u ,say 60^0 with X-axis in this example. Link 3 stretches link 4 straight, however, link 4 is just in its change point position. Suppose link 4 will rotate clockwise if link 2 rotates clockwise. (Fig.I.2 shows the relationships among the angler displacements of links 2, 3, and 4; i.e. abscissa ϕ and curves μ and ψ .) Link 2 reaches its lower limit position when it makes angle $\phi_l (= -60^0)$ with X-axis. Meanwhile, the maximum swing angle of link 4 is about 215^0, the upper half angle $\phi_u \approx 97^0$, the lower half angle $\psi_l \approx 118^0$. Besides, there are two crests ($\mu = 90^0$) and one valley ($\mu \approx 45^0$) in curve μ. Draw a horizontal line passing through the lowest point, $\mu_{min} = 54^0$, of the valley intersecting curve μ in first quadrant at the point, $\mu_1 = 54^0$, where the corresponding position ϕ_1 presents. It is not necessary to find out the another intersecting point where $\phi_2 (\mu_2 = 54^0)$ in second quadrant is, because the crank position where ψ in its extremum, $\phi_{\psi ext}$, is previously reaching than ϕ_2 is. In addition, the designed allowable output oscillating angle $[\psi]$ must be less than $\psi_1 + |\psi_{ext}|$. The allowable transmission angle $[\mu_{min}]$ is defined less than or equal to μ_{min} which is 54^0 in this example.

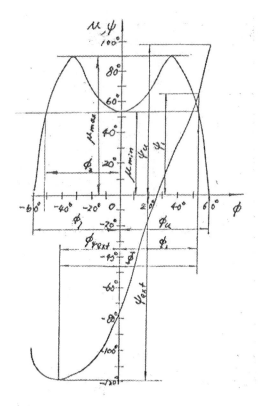

Figure I.2 Curves on displacement of coupler 3 and output link 4
versus that of driving link 2 in a double rocker mechanism

Let $[\mu_{min}]$ be a set of discrete values of $30^0, 45^0$ and 60^0, matching with a series of different relative lengths of links. Then thirteen curves and corresponding data are obtained which are shown in Fig.I.3 and Table I.1 as follows:

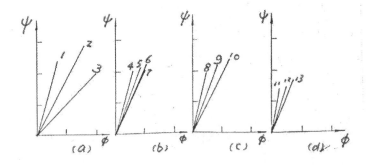

Figure I.3 Theoretical mean ratios between output and input angles in
graph, i.e. $(\psi_1 + |\psi_{ext}|) / (\phi_1 + |\phi_{yext}|) = \Psi^* / \Phi^*$

---

*Ψ is the theoretical maximum output oscillating angular displacement
$(=\psi_1 + |\psi_{ext}|)$, Φ is the corresponding input crank anglar displacement
$(=\phi_1 + |\phi_{yext}|)$

Table I.1 Ψ/Φ Values with Different $[\mu_{min}]$ and Relative Link Lengths

Series	a			b				c			d		
Number	1	2	3	4	5	6	7	8	9	10	11	12	13
L_1	Unity												
$[\mu_{min}]$	30^0			45^0							60^0		
L_2/L_1	1.25	1.5	1.75	1.5				1.25	1.5	1.75	1.25	1.5	1.75
L_2-L_1	$L_3/2$			$1.47 L_4$	L_4	$0.76 L_4$	$0.70 L_4$	L_4			$L_4/\sqrt{3}$		
L_4/L_3	$\sqrt{3}/2$			$1/2$	$\sqrt{2}/2$	1	$\sqrt{2}$	$\sqrt{2}/2$			$1/2$		
$\psi_1+\lvert\psi_{ext}\rvert$	217^0	225^0	180^0	184^0	192.5^0	199^0	184^0	176.5^0	193^0	212^0	126^0	130.5^0	151.7^0
$\phi_1+\lvert\phi_{\psi EXT}\rvert$	67.5^0	142^0	180^0	56.5^0	78.5^0	97^0	91.5^0	39.5^0	77.5^0	116^0	23.5^0	42^0	65^0
Ψ/Φ	3.21	1.79	1	3.25	2.45	2.05	2.01	4.46	2.49	1.82	5.36	3.09	2.33

【Example I.1】 Design a double rocker mechanism with an input angular displacement 90^0 and an output angular displacement 210^0 approximately, $[\mu_{min}]=40^0$.

Solution: Refer to Table I.1, we see that series b no. 6 is close to the demand of this example. Draw a configuration with $1:1.5:\dfrac{0.5}{0.76}:\dfrac{0.5}{0.76}=L_1:L_2:L_3 L_4$, $[\mu_{min}]=40^0$, Fig.I.4 a. We have

$$\Phi = \phi_1+\lvert\phi_{\psi ext}\rvert = 64.5^0 + 48^0 = 112.5^0$$
$$\Psi = \psi_1+\lvert\psi_{ext}\rvert = 86^0 + (71^0 + 53^0 + 9^0) = 219^0$$

Obviously, these values of Φ and Ψ are both greater than needed. Reduce Ψ from 219^0 to 210^0 (i.e. $86^0 + 71^0 + 53^0$, the initial position unchanged), then Φ will reduce to 91^0 (i.e. $64.5^0 + 26.5^0$). These values meet the requirements asked for. The three slant lines of ψ versus ϕ in Fig.I.4b represent the average values in three stages correspondingly.

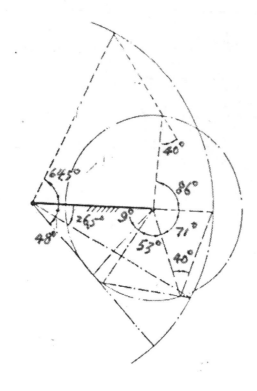

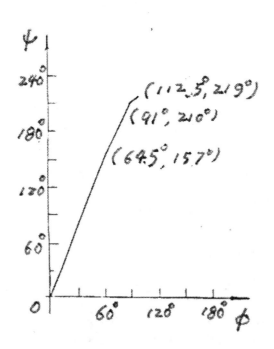

Figure I.4   A double rocker mechanism with ϕ, ψ, $[\mu_{min}]$ equal to 90^0, 210^0, 40^0 respectively: (a) configurations shown in four positions; (b) ψ versus ϕ in three stages.

APPENDIX J   Acceleration Equation for Mechanisms Connected in Series

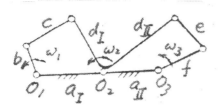

Figure J.1   Mechanisms connected in series

Let $\theta_1, \theta_2, \theta_3$ be the angular displacements of shaft O_1, O_2, O_3; $\omega_1, \omega_2, \omega_3$ be the angular velocities of shaft O_1, O_2, O_3; $\varepsilon_1, \varepsilon_2, \varepsilon$ be the angular accelerations of shaft O_1, O_2, O_3 respectively, and $\omega_1 = $ constant, then

$$\omega_3 = \frac{d\theta_3}{dt} = \frac{d\theta_3}{d\theta_2}\frac{d\theta_2}{d\theta_1}\frac{d\theta_1}{dt} = n_{II}n_I\omega_1$$

where $n_I = \dfrac{d\theta_2}{d\theta_1} = \dfrac{\omega_2}{\omega_1}$

$$n_{II} = \frac{d\theta_3}{d\theta_2} = \frac{\omega_3}{\omega_1}$$

n_I and n_{II} are the transmission ratio, i.e. the reciprocal of velocity ratio, between corresponding shafts separately.

$$\begin{aligned}\varepsilon_3 &= \frac{d\omega_3}{dt} = \omega_1(n_I\frac{dn_{II}}{dt} + n_{II}\frac{dn_I}{dt})\\ &= \omega_1[n_I\frac{dn_{II}}{d\theta_2}\frac{d\theta_2}{d\theta_1}\frac{d\theta_1}{dt} + n_{II}\frac{dn_I}{d\theta_1}\frac{d\theta_1}{dt}]\\ &= \omega_1^2(n_I^2 A_{II} + n_{II}A_I)\end{aligned} \qquad (J.1)$$

where $A_I = \dfrac{dn_I}{d\theta_1} = \dfrac{d^2\theta_2}{d\theta_1^2} = \dfrac{d^2\theta_2}{\omega_1^2 dt^2} = \dfrac{\varepsilon_2}{\omega_1^2}$

$$A_{II} = \frac{dn_{II}}{d\theta_2} = \frac{d^2\theta_3}{d\theta_2^2} = \frac{d(\omega_3/\omega_2)/dt}{d\theta_2/dt} = \frac{\omega_2\varepsilon_3 - \omega_3\varepsilon_2}{\omega_2^3}$$

Let $\dfrac{\varepsilon_3}{\omega_1^2} = A_0$, then

$$A_0 = n_I^2 A_{II} + n_{II}A_I \qquad (J.2)$$

This is the acceleration equation for mechanisms connected in series. A_0 is the specific overall acceleration, A_I and A_{II} are the specific accelerations of the first and second component mechanisms respectively.

APPENDIX K Modified Hain's Output Angular Acceleration Equation of Planar Four-Bar Linkages in Single or in Double Connected in Series

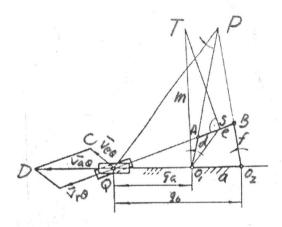

Figure K.1 A basic four-bar linkage with two imaginary pin connected sliders for analyzing angular velocities and angular accelerations of links.

Take the crank rocker mechanism O_1ABO_2 as an example, Fig.K.1. When crank d rotates with a constant angular velocity ω_d, rocker f with an angular velocity ω_f and an angular acceleration ε_f which would be more difficult to find out the ε_f than ω_f with the known configuration as shown. However, it may be easier if the Hain's angular acceleration equation, Reference [10], is used and modified as follows:

Since point Q is the instantaneous center of links d and f where there are equal linear velocities belonging to links d and f respectively, hence

$$\overline{V}_{Qd} = \overline{V}_{Qf}$$
$$\overline{\omega}_d \times \overline{q}_a = \overline{\omega}_f \times \overline{q}_b \qquad (K.1)$$

Where $\overline{q}_a$ and $\overline{q}_b$ are the vector distances from the rotating centers of corresponding links to instantaneous center Q respectively.

Differentiating e.q.(K.1) with respect to time, we have

$$\overline{\varepsilon}_d \times \overline{q}_a + \overline{\omega}_d \times \dot{\overline{q}}_a = \overline{\varepsilon}_f \times \overline{q}_b + \overline{\omega}_f \times \dot{\overline{q}}_b \qquad (K.2)$$

where $\overline{\varepsilon}_d$ and $\overline{\varepsilon}_f$ are the angular accelerations of links d and f, $\dot{\overline{q}}_a$ and $\dot{\overline{q}}_b$ are the increasing or decreasing velocity of vector radii $\overline{q}_a$ and $\overline{q}_b$ respectively. $\dot{\overline{q}}_a$ and $\dot{\overline{q}}_b$ are both equal to the velocity of instantaneous center Q, $\overline{V}_{aQ}$, and closely related with ω_d, i.e.

$$\dot{\overline{q}}_a = \dot{\overline{q}}_b = \overline{V}_{aQ} = \overline{\omega}_d \times \overline{m} \qquad (K.3)$$

where $\dot{\overline{q}}_a$ and $\dot{\overline{q}}_b$ are both equal to the cross product of $\overline{\omega}_d$ and $\overline{m}$ which is effected by $\overline{m}$ if $\overline{\omega}_d$ remains constant. $\overline{m}$ is called the radius of translation, Reference [8], its length unit is the same as that of the links.

The verification of eq.(K.3) is as follows:

Let coupler e (i.e. AB) and frame a (i.e. O_1O_2) both have slider of their own and a common revolute joint at Q. Q becomes a compound pivot joint, its absolute velocity $\overline{V}_{aQ}$

moves along the center line of frame a. Assume point Q is merely a point coincident on coupler e (the moving coordinate system), then the velocity of Q's coincident point on coupler will be

$$\overline{V}_{eQ} = \overline{\omega}_e \times P\overline{Q}$$

However in this instant, point Q is the point moving along the center line of coupler e relatively, so it must have a relative velocity $\overline{V}_{rQ}$ relative to e. The absolute velocity of Q, $\overline{V}_{aQ}$, will be the geometrical sum of $\overline{V}_{eQ}$ and $\overline{V}_{rQ}$, that is

$$\overline{V}_{aQ} = \overline{V}_{eQ} + \overline{V}_{rQ}$$

Draw a line parallel to PQ (i.e. the collineation axis) from point O_1 intersecting with coupler e at S. Through S draw a line perpendicular to coupler e intersecting with the perpendicular of frame a from point O_1 at T, then

$$\Delta O_1 ST \sim \Delta QCD, \text{ because}$$

$$CD \perp ST, QD \perp O_1 T, \ CQ \perp PQ \parallel O_1 S$$

Then $\dfrac{O_1 T}{V_{aQ}} = \dfrac{O_1 S}{V_{eQ}}$

Since $V_A = O_1 A \times \omega_d$, $V_A : V_{eQ} = PA : PQ$

hence $V_{eQ} = (\dfrac{O_1 A \times \omega_d}{PA}) PQ$

Furthermore $\Delta PQA \sim \Delta O_1 SA$, then

$$O_1 S = PQ(\dfrac{O_1 A}{PA})$$

whereas $\dfrac{O_1 T}{V_{aQ}} = [\dfrac{(PQ)(O_1 A)}{PA}] / [\dfrac{(O_1 A)(\omega_d)}{PA}(PQ)]$; we have

$$\overline{V}_{aQ} = \overline{\dot{q}}_a = \overline{\dot{q}}_b = \overline{\omega}_d \times \overline{m} \ (\text{ same as eq.(K.3)})$$

where $O_1 T$ is replaced by m, the whole equation is vectorial so as to have a sense significance as well as magnitude.

Reviewing back to Fig.K.1 , we have

$$\omega_f = \pm \dfrac{q_a}{q_b} \omega_d \qquad\qquad\qquad (K.4)$$

When point Q is outside of $O_1 O_2$ (as in Fig.K.1), eq.(K.4) is positive; inside , negative.

Eq.(K.2) may rewrite as

$$\overline{\varepsilon}_f \times \overline{q}_b = \overline{\omega}_d \times \overline{\dot{q}}_a - (\dfrac{q_a}{q_b})\overline{\omega}_d \times \overline{\dot{q}}_b + \overline{\varepsilon}_d \times \overline{q}_a$$

Premultiplied by $\overline{q}_b$, then

$$\overline{q}_b \times \overline{\varepsilon}_f \times \overline{q}_b = \overline{q}_b \times \overline{\omega}_d \times \overline{\dot{q}}_a - \overline{q}_b \times (\dfrac{q_a}{q_b})\overline{\omega}_d \times \overline{\dot{q}}_b + \overline{q}_b \times \overline{\varepsilon}_d \times \overline{q}_a$$

From eq.(K.3) and $(\dfrac{q_a}{q_b})\overline{q}_b = \pm \overline{q}_a$ (position or negative is determined by the same rule as

eq.(K.4)), we have

$$-q_b{}^2\bar{\varepsilon}_f = -\omega_d{}^2\bar{q}_b \times \overline{m} + \omega_d{}^2\bar{q}_a \times \overline{m} - q_a q_b \bar{\varepsilon}_d$$

$$-\bar{\varepsilon}_f = \frac{-\omega_d{}^2}{q_b{}^2}(\bar{q}_b - \bar{q}_a) \times \overline{m} - \frac{q_a}{q_b}\bar{\varepsilon}_d$$

$$\bar{\varepsilon}_f = \frac{\omega_d{}^2}{q_b{}^2}(\bar{a} \times \overline{m}) + \frac{\omega_f}{\omega_d}\bar{\varepsilon}_d \qquad (K.5)$$

As prescribed above, the input shaft rotates with a uniform speed, i.e. $\omega_d = constant$, $\bar{\varepsilon}_d = 0$, then

$$\bar{\varepsilon}_f = \omega_d{}^2\left(\frac{\bar{a} \times \overline{m}}{q_b{}^2}\right) \qquad (K.6)$$

Now we are going to consider two four-bar linkages connected in series, see Fig.J.1. The angular velocity ω_b of the input link b of the first component mechanism; namely, the input of the composite mechanism is assumed constant, then the angular acceleration $\varepsilon_b = 0$. The angular acceleration of the output link d of the first component mechanism, namely, the input of the second component mechanism

$$\bar{\varepsilon}_d = \omega_b{}^2\left(\frac{\bar{a}_I \times \overline{m}_I}{q_{bI}{}^2}\right) \qquad (K.7)$$

where subscripts d and b of ε and ω replace f and d respectively, I's are added to a, m, and q_b from eq.(K.6), for all the relationships are among the first component mechanism. As $\bar{\varepsilon}_d$ and $\bar{\omega}_d$ can be easily calculated , the output angular acceleration we have from eqs.(K.5) and (K.7) is

$$\bar{\varepsilon}_f = \omega_d{}^2\left(\frac{\bar{a}_{II} \times \overline{m}_{II}}{q_{bII}{}^2}\right) + \omega_b{}^2\frac{\omega_f}{\omega_d}\left(\frac{\bar{a}_I \times \overline{m}_I}{q_{bI}{}^2}\right)$$

where II is presented for second component mechanism.
Multiply it by $1/\omega_b{}^2$, yields

$$\frac{\bar{\varepsilon}_f}{\omega_b{}^2} = \frac{\omega_d{}^2}{\omega_b{}^2}\left(\frac{\bar{a}_{II} \times \overline{m}_{II}}{q_{bII}{}^2}\right) + \frac{\omega_f}{\omega_d}\left(\frac{\bar{a}_I \times \overline{m}_I}{q_{bI}{}^2}\right)$$

Let $A_0 = \dfrac{\varepsilon_f}{\omega_b{}^2}, n_I = \dfrac{\omega_d}{\omega_b}, n_{II} = \dfrac{\omega_f}{\omega_d}, A_I = \dfrac{\bar{a}_I \times \overline{m}_I}{q_{bI}{}^2}, A_{II} = \dfrac{\bar{a}_{II} \times \overline{m}_{II}}{q_{bII}{}^2}$,

therefore

$$A_0 = n_I{}^2 A_{II} + n_{II} A_I \qquad (K.8)$$

This is the dimensionless acceleration equation for mechanisms connected in series, the same as listed in Appendix J.

Simple Conclusion

(1) The direction of vector $\overline{m}$ is always perpendicular to the frame base line and toward the driving shaft center, whose rotating tendency about the output shaft center is the very sense of angular acceleration of output link.

(2) The velocities of the coincident points of links d and f at Q are $\overline{\omega}_d \times \overline{q}_a$ and $\overline{\omega}_f \times \overline{q}_b$ respectively. Although they are equal each other, they are both different from the linear velocity of instantaneous center Q, i.e. $\overline{V}_{aQ}(= \overline{\omega}_d \times \overline{m})$.

(3) Since m is the function of relative length ratio and position of the mechanism, hence the output angular acceleration ε_f is the function of that of the mechanism too, for ε_f is the function of m.

APPENDIX L   Pressure Angle

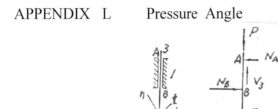

Figure L.1   A typical cam mechanism: (a) configuration;(b) the static force analysis of the knife-edge follower

The big pressure angle is seemed as the unlucky thing in the academic circles generally. The typical example is the pressure angle shown in cam mechanisma,Fig.L.1. According to conventional force analysis, the pressure angle α is the very angle between he velocity $\overline{V}_c$ on the point C of follower 3 and force $\overline{F}_{23}$ where force is being acted from power source cam 2. The vertical component force $F_{23}\cos\alpha$ of force $\overline{F}_{23}$ is the effective force against the load P, the horizontal component force $F_{23}\sin\alpha$ is a force which balances bearing forces N_A, N_B. Whereas a bending moment may cause the rupture of link 3 due to the present of these horizontal forces. Furthermore, the frictional forces on the follower may be the resistance of motion if friction is considered. Obviously the greater the pressure angle α, the less the $F_{23}\cos\alpha$, the more the $F_{23}\sin\alpha$, disadvantages are clear

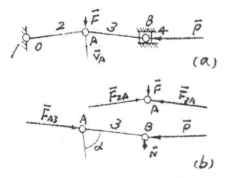

Figure L.2   Toggle mechanisms : (a) configuration;(b) static force analysis

Again, we know hat the toggle mechanism shown in Fig.L.2 is a force amplified mechanism.   The more the pressure angle α on link 3 presents, the more the force $\overline{F}_{A3}$ that link 3 obtains   under the applying force $\overline{F}$. $\overline{F}_{A3}$ will be infinity when α reaches 90^0 theoretically. No doubt, the pressure angle α is welcome on toggle mechanisms.
In view of instantaneous power, we obtain

$$\text{Power} = F_{A3}V_A\cos\alpha$$

Assuming the power of link 3 and the magnitude of V_A keep unchanged in transmission, then the more the α is the more the $\overline{F}_{A3}$. The pressure angle plays a quite different role here.

Finally a planar four-bar linkage shown in Fig.L.3 will be discussed. The pressure angles α_A, α_B belong to links 3 an 4 respectively.

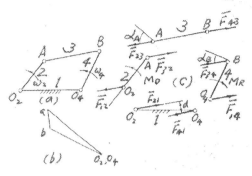

Figure L.3　Planar four-bar linkage(a) configuration;(b) velocity diagram;(c) static force analysis

$M_D + M_R = F_{21} \times d$, where

M_D____driving moment

M_R____working resistance moment

The α_B,pressure angle for the consume power, is the function of $\cos\alpha_B\left(\because P = F_{34}V_B \cos\alpha_B\right)$. The α_B plays an important role here. Though α_A is merely a pressure angle on a member of the transmission chain; which still be discussed for the force analysis of the mechanism. M_D, M_R are the turning moments of input link 2 and output link 4 respectively; the former is the power source, the latter is the power consumer. The moments applied to he frame are M_D, M_R; their sum is equal to $F_{21} \times d$.

260

References

1.Arneson,L., "Planetary-Genevas", *Machine Design*, Sep.1959,pp.135-139.

2.Aronson,R., "Large-Oscillation Mechanisms", *Machine Design*, Nov.1960,pp190-196

3..Bocher,M., *Introduction to Higher Algebra*, Macmillan, New York,194,pp.198-199

4.Chironis,N.P., "New Design and Equations for Gear-Slider Mechanisms", *Product Engineering,*Feb.1965.

5.Fenton,R.G., "Geneva Mechanisms Connected in Series", *J. Of Engineering for Industry, Trans. ASME, Series B*,Vol.97,May 1975, pp.603-608

6.Hain,K., *Applied Kinematics*, McGraw-Hill, New York,1967

7.Hain, k., "The Production of Large Oscillating Angles by Cam-Linkage Mechanism*s"*, *J. of Mechanisms,* Vol. 6,1971

8.Hain, K., "Optimization of Cam Mechanisms——to give good transmissibility, maximal output angle of swing and minimal acceleration", *J. of Mechanisms*, Vol. 6,1971,pp.419-434.

9.Hain,K., "Geared Four-Bar Linkages", *Machine Design*, Vol.34, Oct.11,1962,pp.195-199.

10 Hain,K.,"Winkelbeschlewnigungen sverh*ä*ltnisse in vierbgliedrigen Getrieben", *Werkstat tund Betrieh,* Vol.98, 1965, pp.504-508

11.Ham and Crane, *Mechanics of Machinery,* John Willy and Sons Inc., New York,1955.

12 Harding, B.L., "Hesitation", *J of Engineering for Industry, Trans. ASME, Series B*, Vol.87, May 1965,pp. 205-212.

13.Hartenberg, R.S., and J Denavit, *Kinematic Synthesis of Linkages*, McGraw-Hill, New York , 1964.

14 Hirschhorn,J., *Kinematics and Dynamics of Plane Mechanisms*, McGraw-Hill, new york ,1962.

15.Jenson,P.W.,"Analysis and synthesis of Some Gear Drives, Having as a Basic Mechanisms the Centric Slider-Crank or the Turning-Bolck Linkage",*Trans.Mech.,64-Mech-31*,1964.

16.Kaplan,J., and H.Korth, "Cyclic-Three Drives", *Machine Design*, March 1959,pp.185-188

17Lee,T.W., and Y. Shereshevsky "Kinematic Synthesis of Planar Two-Gear Drives with Prescribed Dwell Charateristics", *J. of Mechanical Design*, Vol.104, Oct. 1982,pp.687-697.

18.Mabie, H.H., and F.W. Ocvirk, *Mechanisms and Dynamics of Machinery"*, John Willy & Sons, New York,1978

19 Martin,G.H.,*Kinematics and Dynamics of Machines,* McGraw-Hill, New York, 1982.

20 Qiu,J.X., and Meng, F.Z., "Analysis and Synthesis of Chained Linkages",(in Chinese) *The third National Seminar of Mechanisms,* August,1985.

21 Rao,A.V.M., and G.N. Sandor, "Extension of Freudenstein's Equation to Geared Linkages", *J. of Engineering for Industry, Trans.* ASME, Series B Vol.93, No.1, Feb.1971 pp.201-211.

22. Rooney,G., "Geared Linkages", *Engineering*, Vol.218, No.9, Sep.1978, pp.880-884.

23 Rose,S.E., "Five-bar Loop Synthesis", *machine design*,Oct.12,1961,pp.189-195.

24 Sender,G.N., and A.G.Erdman, *Advanced Mechanism Design: Analysis and Synthesis,* Prentice-Hall, Inc.,1984.

25Sandor,G.N., R.E. Kaufman,A.G. Erdman ...., "Kinematic Synthesis of Geared Linkages", *J. of Mechanisms,*Vol.5,1970,pp.59-87.

26 Schmidt,E.H., "Cycloidal-Crank Mechanisms", *Machine Design,*Apr.2,1959,pp111-114.

27 Schmidt,E.H., "Cyclic Variations in Speed", *Machine Design,*Mar.1947,pp.108-111.

28 Soni,A.H., *Mechanism Synthesis and Analysis,* McGraw-Hill, New York,1974.

29 Talbourdet,G.J., "Intermittent Mechanisms", *Machine Design,*Sep.,Oct..,Nov.1948

30 Tao,D.C., and A.S. Hall,"Analysis of a Symmetrical Five-Bar Linkage", *Prodution Engineering,*Jan.1053,pp.175-177,201,203,205.

31Tsai,Y.C., and H.H.Soni, "Accessible Region and Synthesis of Robot Arms", *J. of Mechanical Design,*Oct.1981,Vol.103.

32 Vadasz,A.F., and A.H.Soni, "Gross Motion Analysis of Planar Linkages with Application to a Geared Five-Bar", *J. of Engineering for Industry, Trans. ASME, Series B,*Vol.99,Aug.1977,pp674-677.

33. Yan,H.S., and Wu,L.L., "On the Dead-Center Positions of Planar Linkage Mechanisms", *J. of Mechanisms.Transmission, and Automation in Design, Trans. ASME,*1989,pp40-46.

34 Yan,Zhen-Ying, "Kinematic and Dynamic Analysis of 6-link Coupling"(in Chinese),*J. of Wuhan Institute of Technology,*Vol.9,No.3,1987,pp40-48

35 Yen,Zhen-Ying, "Maximum Pressure Angle Analysis of Disk Cam with an Inline Knife Edge Follower"(in Chinese),*Machine Design and Research,*No.6,1987,pp49-53.

36. Yan,Zhen-Ying, "Primary Analysis of 6-link Coupling"(in Chinese),*JIXIE SHELI(Machine Design of China)* No.4,1988,pp29-34

37 Yan,Zhen-Ying, :"Coupler Curves of Five-Bar Linkage of Two-Gear Drives"(in Chinese),*J.of Wuhan Institute of Technology,*Vol.10,No.3,1988,pp44-49

38 Yan, Zhen-Ying, "Graphical Synthesis on the Extension of Output Oscillating Angle",(in Chinese)*SHIYONG JIXIE JISHU (Practical Technology on Machinery),*July 1990,pp.15-18.

39. Yan,Zhen-Ying, "Investigation on Hain's Specific Angular acceleration Equation",(in Chinese) *SHIYONG JIXIE JISHU (Practical Technology on Machinery),*Dec.1990,pp4-7.

40. Yan,Zhen-Ying, "Determination of the Dead-Center Positions of Composite Mechanisms by Instantaneous Centers",(in Chinese),*J.of Wuhan Institute of Technology,*Vol.13,No.3,1991,pp45-51

41 Yokoyama,Y.,"Studies on the Geared Linkage Mechanisms",*Bulletin of the JSME,*Vol.17, No.112. Oct.1994,pp.1332-1339.